Contents

Acknowledgments

After writing this book I owe lots of people favours. My first thanks go to those interviewees who are quoted throughout. All were more than generous with their knowledge and their time. Hats off to: Gillian Appleton, Matthew Arnison, John Arquilla, Scott Balson, John Breen, Mike Childs, Dorothy Denning, Ricardo Dominguez, Esther Dyson, Zack Exley, Matthew Fuller, Lyn Gerry, Frank Guerrero, Jessy from McSpotlight, Mitch Kapor, Gabrielle Kuiper, Steve Kurtz/Critical Art Ensemble, Matthew Lasar, Kalle Lasn, Eric Lee, Edward Liu, Geert Lovink, Veran Matic, Richard Metzger, Dave Morris, Dick Morris, Howard Rheingold, Bruce Sterling, Joe Tan, Dan Tse and Stefan Wray.

Many other people also helped to get this book out. Some read drafts or sent links or made introductions. Others offered comments or encouragement or inspiration or just plain help. Still others bought the drinks. My thanks to: Ien Ang, Franny Armstrong, Kathe Boehringer, Stella Collier and Carla Theunissen, John Finlay, Vlad Keremidschieff, Willa McDonald, Johnnie McLaughlin, George and Pat Meikle, Jimmy and Chrissie Meikle, Kathryn Millard, Tony Moore & all at Pluto, Rick Morgan, Andrew Murphie, Matt Pearce, Peter Phillips, Jenny Pickerill, Rob Pullan, David Ronfeldt, Michael Scott and Jorge Chamizo, Tam, Lynda, Tommy and Michael Sharp, Sarah Shrubb, Mark Thomas, Ray Thomas, Mandy and Afreka Thomson, Paul Thomson, Darren Tofts, Marcus Westbury, Helen Wilson and Sherman Young.

Special thanks to McKenzie Wark for his many helpful suggestions – not least for suggesting I write this book in the first place.

And thanks, most of all, to Fiona Taylor, without whom the book would not, and could not, have been written.

Any flaws in *Future Active* are, of course, down to me; any good points it might have would not have been possible without the people above.

Graham Meikle
Sydney, March 2002

Preface

Graham Meikle's *Future Active* is a wonderfully readable account of what is still a very new series of developments in the use of the Internet by political activists. This book uses original interviews with participants in the proliferating forms of Internet activism. There is solid primary research here which will make the book a standard reference point.

Meikle proposes a taxonomy of forms of Internet activism, and this is a major contribution. Many kinds of activism are currently discussed under vague general headings, as if they were much the same thing. Meikle shows why, for example, 'culture jamming' is not the same as 'hacktivism' – and gives an excellent explanation of both.

Meikle distinguishes between interactive media and what he calls 'unfinished' media. The former offer choices within a market-based notion that one is choosing among equivalents; the latter proposes a genuinely unfinished future that people have to make for themselves. He begins a powerful critique of the limits of utopian thinking about interactive media by directly relating the phenomena of interaction with the question of what it is that one interacts with and for.

Interactivity assumes that the 'user' follows predetermined pathways to conclusions decided in advance ('finished'), and that media is just about commodification, about choice among equivalents. The unfinished, on the other hand, implies an unknown future, one we head into forwards rather than backwards, as it won't necessarily look like the past. It won't be *more* choice, it will be *different* choice.

Meikle links 'unfinished' to the idea of the 'open', as in open source, and the openness of the Indymedia movement. The story that matters is the unfinished, open-ended one in which people collectively make up their future as they go towards it, facing it, as consciously as possible. The narrative strategies of mainstream media are closed, not open, and presume a limited menu of points of view among which one must be 'balanced', but all of which see the future in the same way.

Thus, crucially, making different kinds of futures means making different kinds of media. Media that, as Meikle suggests, also includes the nasty right-wing populist options. An innovation of this book is that it considers right-wing Internet activism and makes useful and interesting comparisons between what the left and the right are doing out there in the Internet. It opens up a line of inquiry too often ignored.

The open construction of the future means confronting these bad desires head-on in an open media world, rather than suppressing or censoring them, or merely being distracted by the consumerist spectacle. In other words, the absence of ratbag views in mainstream media is not a good thing, as it circumvents the open struggle to make meaning and make history. It is only by creating open media spaces that people can come to make their own futures. And even a messy home-made future will be better than the McWorld of 'interactivity'.

McKenzie Wark
State University of New York

Introduction

First we were told we could all make a difference. Then we were told we could all make a killing. Not so long ago, a new communications technology began to draw a lot of attention. It had already been around for a number of years, with military investment fuelling its early development. But it had since gone through wider experimentation and adaptation, until tens of thousands of ordinary people were using it to connect with each other about all kinds of interests. Corporations, of course, spotted this, and began to search for ways to make it profitable. Yet just as this commercialisation process began, we also began to hear a lot about the potential the medium had for democracy.

Here are some things that were said. The new medium could be 'the most wonderful public communication system imaginable, a gigantic system of channels ... capable not only of transmitting but of receiving, not of isolating [the user], but connecting [the user]'.[1] Users of this technology could 'leap around the world [and wipe] out for all time the age-old barriers of race and language and distance'.[2] It would give the public access to information to let us see through the rhetorical tricks of politicians. So government would become 'a living thing to its citizens',[3] and this would give us 'a new kind of statesman and a new kind of voter'.[4]

That technology was, of course, radio. All those quotes are from the 1920s and 1930s, long before talkback shock-jocks or high-rotation classic rock stations. There was a lot of debate about the democratic potential of radio, while initially there was little understanding of its commercial possibilities. But as the advertising-driven commercial model emerged, together with the beginnings of large-scale networks, corporate interests were able to lobby for the scarce public resource of the airwaves to be handed over to them, and all with hardly any political or public debate.[5]

Seventy years later, the parallels with the Internet are striking.[6] A potentially radical new technology has become the site of confusion about who should control it, and to what ends. What's clear is that much of the rhetoric about cyberspace echoes what's been said before – the postmodern era was supposed to be marked by a loss of faith in grand narratives like progress or emancipation,[7] but the people who write the press releases have yet to hear that one. Dot.com shopping sites talk of 'community', opinion pollsters about 'empowerment'. Our new machines will set us all free.

Just as with radio, when the established media first picked up on the Internet, we heard a lot about its democratic potential. Even as they took the first clumsy steps towards regulating cyberspace, politicians such as Al Gore spoke of how it would herald 'a new Athenian age of democracy'.[8] Journalists, understandably fascinated by the power of media, followed suit – typical was the *New York Times* in 1995, announcing on its front page that 'Anyone with a modem is potentially a global pamphleteer.'[9] Reporters took cues from best-selling pundits such as Nicholas Negroponte and Howard Rheingold. MIT academic Negroponte, for instance, predicted that 'many of the values of the nation-state will give way to those of both larger and smaller electronic communities'.[10] In *The Virtual Community*, still one of the most influential books about the Net, Rheingold wrote of its potential to 'revitalise citizen-based democracy', sketching a world in which 'every citizen can broadcast to every other citizen'.[11] The revolution would not be televised. It was going to be online instead.

But readers of the corporate press releases that come disguised as newspaper computer sections will have detected a pronounced shift in this rhetoric – the wonders of e-commerce soon became the message, as the electronic *agora* began to get paved over and turned into the virtual mall. Even though the fall of the NASDAQ and the dot.com shakeout validate novelist William Gibson's sci-fi definition of cyberspace as a 'consensual hallucination',[12] the popular debates about the Net still concern how to get – and stay – rich quick. And, of course, how to spend up big. A browse through the archives of *Wired* magazine illustrates this. Sift through early issues and we find the high-profile libertarians of the Electronic Frontier Foundation making the case for 'a Jeffersonian information policy',[13] or outlining what readers could do to influence the US government on digital privacy.[14] Click forward to the March 2000 issue, which has a giant dollar sign on its cover, and we're greeted with breathless headlines promising to reveal 'How you'll buy and sell in the agent-driven, super-fluid, universal marketplace', and to introduce us to 'kick-ass shopbots'.[15]

Politics.net

Future Active tests the early claims for the democratic potential of the Internet. So this is a book about Net politics – but what kinds of politics? Political scientist David Resnick draws a useful distinction between three forms of Internet politics: 'politics within the Net, politics which impacts the Net, and political uses of the Net'.[16]

The first of these describes the internal politics of Net communities, the social dynamics of text-based interaction, through which participants create a group identity and police the boundaries of that group. These are the micro-politics of email, listservs and discussion forums, more than of the web. These are the Net politics of free software and of the hacker ethic that 'information wants to be free'. They relate to the early cyberculture which developed netiquette, and worked out ways of coping with phenomena such as flaming (the exchange of hostile messages) and spam (commercial junk email).

Resnick's second category relates to the real-world politics which affect the Net – issues of access, ownership and control, of censorship and regulation – as cyberspace is brought within the boundaries of wider social environments. These real-world politics would include such examples as the content controls introduced in Australia through the *Broadcasting Services Amendment (Online Services) Act*

1999, and the intellectual property wrangles which strangled Napster.

Resnick's third category refers to political uses of networked computers that attempt to effect social or cultural change in the *offline* world: a politics which can use the graphical and multimedia capabilities of the web as well as text-based applications such as email. *Future Active* is about this third form of Internet politics – Internet activism.[17]

We'll examine various projects that use the Net to try to effect social, cultural or political change – from hackers, culture jammers and corporate saboteurs to established political parties. It's a huge terrain – there are Net campaigns around the world and around the corner. So this book makes no attempt to be *comprehensive* in its coverage – that would require a map much larger than the territory itself. Instead, it looks at some *representative* approaches to online politics, and how these relate to the established media. I don't intend this book to be the last word on the subject, but to be a contribution to an ongoing conversation that I hope many others will join.

A starting assumption is that it no longer makes sense – if it ever did – to treat the Net as an entirely separate realm. Instead, we have to see it as part of the broader media environment. There's a sense in which I may be writing at a moment in which the Internet is *over* – it may be that the technology is already so widespread, so developed, and so fully integrated into the mediascape that it makes no sense to write a book about the political uses of the Net in *isolation*. It's already beginning to sound awkward, for instance, to talk about 'the Internet community'. That 'community' is already too big, too diverse and diffuse, too ubiquitous. But it doesn't yet sound as awkward as it would to talk about 'the telephone community'. Until it does, then there's still a space – and a need – to try to think through some of our relationships to new media.

To that end, I've tried to relate the projects examined in this book to their wider media environments. The Internet is an increasingly crucial part of broader media activism – a new tool in the box, but not the only one. All the groups and campaigns in this book rely on the Net, but they also rely on the more established media. In some cases they define themselves in opposition to those media; in others, they use the Net to try to force their way onto the wider media agenda. As we'll see, people who hope to draw attention to issues can use the Net in a host of ways, but few are effective without the eventual participation of the older media.

Coming up

In the following chapters we'll look at a range of Internet campaigns from three perspectives. First, we'll consider the nature and potential of *interactivity*. Second, we'll examine the traditions and characteristics of *alternative media*. And third, we'll introduce the emerging models of *tactical media*. The six main chapters of the book come in pairs: chapter 1 introduces a simple model for thinking about the Internet's potential. Chapter 2 uses that model as a way into evaluating the early hype about the Net and social change.

In chapter 3 we'll explore some high-profile Net campaigns from the perspective of alternative media – by which I mean media that are independent; in some senses dissident (or dissonant); and focused on horizontal organisation, in contrast to the top-down approach of the established media. We'll focus on the Net use of Belgrade radio station B92 during the Milosevic regime and the NATO air strikes; and on McSpotlight, the online support movement for the defendants in the UK's McLibel trial. In chapter 4 we'll build on these ideas of alternative media in an analysis of the global Indymedia movement: a network of websites committed to open publishing and to flattening the vertical hierarchies of established media.

In chapter 5 we'll look at tactical media – at hit-and-run guerilla-media campaigns, at culture jamming, and at online sabotage. We'll look at campaigns from Buy Nothing Day to gwbush.com, the website that provoked the US President to declare that 'there ought to be limits to freedom'. And in chapter 6 we'll develop the tactical approach in an examination of hacktivism and electronic civil disobedience. The 'alternative' and 'tactical' labels are not mutually exclusive, by the way, but can complement each other. As we'll see, some Net campaigns – such as that in support of Radio B92 – illustrate key features of both. The terms are ways of opening things up for analysis and discussion, rather than neat taxonomical boxes. In the epilogue, I give the last word to the people who are interviewed throughout, asking them to offer advice to others who want to use the Net in similar ways.

But we'll begin with an event from late in 1999, when the world had its most vivid demonstration to date that Internet activists are not to be ignored.

CHAPTER 1

Backing Into The Future

Version 1.0

30 November 1999: There's a riot going on. The National Guard move in to enforce a curfew as a state of emergency is declared in Seattle. The dot.com capital, home to Microsoft and Amazon, becomes a battlefield as the World Trade Organisation (WTO) meeting descends into a mixture of high drama and low farce, and tear gas is turned on women dressed as turtles. Proceedings are halted for a while as delegates are kept out by the most eclectic coalition imaginable – Christian punks, Zapatistas, steelworkers unions, church councils, Greenpeace, Tibetan monks, the Biotic Baking Brigade, the Anarchists Soccer League, the Association of Autonomous Astronauts. But what's significant about the 30 November (N30) protesters, who are not confined to Seattle, is more than just their global opposition to globalisation – it is their use of the Internet to organise, publicise and mobilise.

Groups preparing for Seattle did much of their work on the Net. Established organisations – the Sierra Club, Corporate Watch, unions – were joined by new coalitions, such as the Direct Action Network, which grew from three groups to 70 in the months leading up to N30. Websites offered resources to those planning to attend, from maps of Seattle to legal advice, while email and listservs were used to coordinate, inform, organise and train. The Ruckus Society used the Net to provide manuals and organise action camps for training in nonviolent civil disobedience, from urban abseiling to crafting soundbites. And as well as influencing the offline events, cyberspace was also the actual site of some anti-WTO action. Corporate sabotage specialists ®™ark (pronounced 'art-mark') created a sophisticated parody of the WTO's website to provide counter-spin, while the Electrohippies mounted a virtual sit-in of the real WTO site.[1] And when events began, the online Independent Media Centre uploaded real-time reports from activists in Seattle, and streamed audio and video coverage.[2]

Rubber bullets in downtown Seattle were probably not the kind of thing that early proponents of Internet democracy had in mind. But Seattle and the similar events that followed in Washington, London, Melbourne, Prague and Genoa are the most dramatic manifestations of global Net activism – not, perhaps, the most successful, but certainly the most vivid example of a possible Internet future Version 1.0. It may not be the new Athenian age of democracy that Al Gore was on about, but it is, for better or worse, the politics of the Internet in action. I call it the Internet Version 1.0 because it connects to those early claims that the Net would bring about huge changes in political life and social action. Version 1.0 is the Internet as an *open* system. So it doesn't apply only to early applications such as Usenet, but also to more recent developments which assume and enable an open system: file-sharing applications such as Napster or Gnutella, for instance.

The battle of Seattle was a huge media event – finally a sexy angle on trade talks! – which in itself points to one of the key strengths of Net-based politics: one way to measure the success of many of the projects in this book is to ask how effectively they can use the Net to force their cause onto the agenda of the mainstream media. If the Seattle demonstrations were any kind of success, it was in dragging globalisation off the business pages and into the headlines. The extent to which the Seattle protests hijacked the media agenda of the WTO meeting was a more significant achievement than keeping Kofi Annan in his hotel room for an hour or trashing a Starbucks.

But forcing the older media to pay attention is one thing. Telling them how to do it is another. Writing about Seattle, pro-globalisation columnists and economists queued up to tell us how the protesters just didn't get it. Like the Borg with MBAs, they said resistance was futile. The protesters hadn't spent enough time poring over the balance sheets; they just didn't see the benefits. Here's a small sample – those opposed to globalisation were 'economic Luddites',[3] 'flat-earth advocates' and 'yuppies looking for their 1960s fix',[4] who believed in nothing more substantial than 'the latest incarnation of a protest ideology'.[5] In such an analysis, in which progress is measured by growth, and both of these are unqualified good things, globalisation is defined surprisingly narrowly – as the expansion of free market ideologies and practices.[6]

But other perspectives are possible, and it is these other perspectives that underpin many of the campaigns and projects in this book. Perhaps the chaos of Seattle, with all its obvious contradictions and conflicts of interest between the groups involved, *is what globalisation is.* Perhaps it's not just that the meaning of the word 'globalisation' itself is contested, but that the cultural, social and economic processes it describes are fundamentally about contestation; about friction, disjuncture, contradiction, difference – a point we'll come back to in chapter 5.

This virtual phase of the battle of Seattle, then, in which the economists were the ones throwing the rocks, points to an initial challenge for Internet activists: how to find ways of using – and working with – the established media without being misrepresented or marginalised. And that's a major challenge, because there is, of course, also an Internet future Version 2.0.[7]

Version 2.0

> Will the highways on the Internet become more few?
>
> George W. Bush[8]

10 January 2000: America Online (AOL) and Time–Warner merge in the biggest deal in corporate history, with the new company becoming a larger economic entity than Coca-Cola or Brazil.[9] In a neat illustration of media convergence, The Australian*'s front-page photo of Time–Warner's Gerald Levin and AOL's Steve Case comes framed in a* *desktop window graphic, complete with scroll bar and cursor. As a first step, AOL will offer its services to Time–Warner cable subscribers via a set-top box. Whether you're listening to REM, watching* Buffy, *renting* The Matrix, *reading* Sports Illustrated, *checking out CNN or surfing with Netscape, the new company will be getting paid. CNN founder Ted Turner declares that this is all more exciting than sex, while, at the height of the hype, search engine Yahoo! is touted as a possible buyer of News Corporation or Disney – a notion put into perspective by satirical online newspaper* The Onion, *which reports that the website blairwitchproject.com is going to buy General Motors.*

Like other kinds of stories, of course, news stories need narrative drive, need action, characters, conflict.[10] So the event is given greater emphasis than the process; the personalities of Case, Levin and Turner are highlighted; and there is much speculation about possible friction between them – can Levin work with Case? Will Turner be happy as number three? – and so on and on. But in terms of the Internet's democratic potential, the real story is summed up by one *Sydney Morning Herald* journalist who visits AOL Australia's site to ask users what they think of the new merger, only to be booted out by a chat-room bouncer who tells him 'that sort of thing is not allowed on AOL'.[11]

This is the future of the Internet Version 2.0, in which we get to watch TV and type at the same time. It's the Net as a *closed* system, rather than an *open* one. A closed-system Internet is the e-commerce holy grail, and for a time it seemed that every search engine and service provider was trying to turn itself into a portal, or one-stop Net shop, closing the system as much as it could. Like other mega-portals, AOL aims to corral its users into pre-selected sites from which AOL generates advertising revenue. The idea is that when subscribers log on to check a sports result or book a plane ticket, they'll allow themselves to be steered through a 'channel' of participating sites, with every step of the search exposing them to more ads. Portal sites hope we will keep their home page set as the default and will never see the need to surf outside the portal's lucrative, closed system. And it often works – AOL reportedly manages to contain its users within this fenced-off area for more than 80% of their time online,[12] while one ratings company reports that in December 2001, eight of the top ten most visited sites in Australia were large commercial portals.[13] The obvious next step here is to charge for access – key sites, from the *Financial Times* to Slashdot, from Yahoo! through Hotmail to the Internet Movie Database, continue to experiment with models of charging for certain kinds of services.

In setting up this opposition between Versions 1.0 and 2.0 I don't mean to over-simplify things. There are tensions between these models of open and closed systems which are important to note. One project that illustrates such tensions is The Hunger Site. This website shows a world map on which one country after another turns black – each time this happens it signifies that someone has starved to death in that country.[14] Visitors can click on a button which takes them to a page of ads. In exchange for viewing these, the site's sponsors pay for a donation of a serving of wheat, rice or maize. After one year of operation, The Hunger Site had recorded more than two million donations; another year after that, donations had reached 198 million.

In some senses, The Hunger Site is a Version 1.0 project. It encourages a limited form of action and promotes open participation. In fact, so many people have promoted it by forwarding the address or adding it to their email signature that its owners have actually asked supporters to stop, as it's becoming counter-productive. For John Breen, who created the site in June 1999,[15] the project was about education and communication.

'The more that corporations are global in scope,' says Breen, 'and the more that international business grows in each developing country, the more various people will communicate with each other all over the world and realise that we are all very similar human beings. All this, combined with the increasing information and education and discussion available to everyone via the Internet, you would think would promote "good" in the world.'

But it's also very much part of Version 2.0 – for one thing, it's *not* a non-profit organisation; for another, it recasts active participation in a problem as simply a matter of viewing more ads. But for its founder, the site was also a way of promoting corporate responsibility.

'With increasing information about corporations available to everybody globally,' Breen says, 'it increases the likelihood that they are going to try to do "good" things, however that is defined by them and the people they are trying to attract. It is kind of the same argument that "bad" dictators can't get away with as much any more in an increasingly informational world, only that it is rewarding "good" rather than punishing "bad".'

There are other tensions between Versions 1.0 and 2.0. Commercial influences, for example, had a key role in enabling the development of a Version 1.0 open Net. If the initial Net infrastructure was built on government subsidies, its spectacular growth in the mid-1990s was in part fuelled by hidden *corporate* subsidies – the free distribution of web browsers; free email services; and the provision of free server space for personal websites by companies such as Geocities (which has since been bought by Yahoo! and is moving towards a subscription model).[16] And at the same time, some very Version 2.0 projects have aspects that illuminate Version 1.0. Matthew Arnison of Sydney Indymedia – whom we'll meet properly in chapter 4 – points out that an important part of the appeal of Amazon.com is a very Version 1.0 idea: *open publishing*, or enabling anyone to supply the content. 'The big thing about Amazon,' he says, 'is the community of book reviews it fosters – Amazon has a strand of open publishing in its heart.' So by using the Versions 1.0 and 2.0 model, rather than trying to over-simplify, I'm proposing two broad poles which we can navigate between.

To some readers, the label 'Version 2.0' might suggest an improvement – more features, fewer bugs. That's *not* how I'm using it. It'll certainly be *sold* to us that way, but Version 2.0 is, for me, the lesser option. In the sense in which I use the terms, Version 1.0 offers change; Version 2.0 offers more of the same. Version 1.0 demands openness, possibility, debate; Version 2.0 offers one-way information flows and a single option presented as 'choice'. Version 1.0 would try to bring the new space of virtual possibility into the world as we know it; Version 2.0 would take the world as we know it – politics-as-usual, the media-as-before, ever more shopping – and impose it upon cyberspace. Version 1.0 would open things up. Version 2.0 would nail them down.

The tensions between open and closed systems are central to the campaigns in this book – as indeed they're central to the development of the Net itself. It's a media myth that the original ARPAnet network was intended to be a Cold War failsafe system to preserve communications in the event of nuclear attack. In fact, it was developed to enable scientists and other academics to share resources in the days when computers were hugely expensive, room-filling behemoths. The real impetus was resource *sharing*.[17] Openness was built into the network from the beginning, and many subsequent key developments – from email to Usenet, from the PC modem to bulletin board systems, from the early web to Linux – have built on the desire to share tools and information. 'My vision,' says Tim Berners-Lee, creator of the world wide web, 'was a system in which sharing what you knew or thought should be as easy as learning what someone else knew.'[18]

As we'll see in the next chapter, the early claims for cyberdemocracy assumed an open system. And attempts to create or sustain openness of one kind or another, from open publishing to open source software, are central to much Net activism – we'll look, for example, at the network of Independent Media Centres set up in the wake of Seattle, which rely for their very existence on open participation and access; and we'll look at politically motivated hackers – hacktivists – who draw on the strengths of the open system and the vulnerabilities of a closed one. But as the (il)logic of commercialisation tends towards closed systems, we'll also look at some projects which engage with commercialisation and global corporations, including culture jammers, anti-corporate saboteurs and the McSpotlight campaign.

So does it work?

There's a question that can't wait until the end of the book – because it's the one everybody always asks first: *how effective is it?* Does Internet activism 'work'? The answer, of course, depends on what people are trying to achieve. So the question is better framed as: in any given case, does the Internet help people to accomplish the given goal? And, often as important as the given goal, what are the unintended or unforeseen outcomes? The next part of this chapter looks at three examples of Internet activism which 'worked' in one way or another. These examples also illustrate some of the key tactics used by Net activists more broadly, so they're followed by a brief overview and summary of some typical approaches to political Internet use.

Each of these three examples illustrates a Version 1.0 approach: each bypasses the established media and relies on horizontal connections among participants. Each represents a grassroots, bottom-up approach to communication and action. And yet each also relies, to some extent, on the participation of the established media. This is not a bug, but a feature – finding ways to exploit the older media on the way to building new ones is fundamental for new media activists. One more thing to note here is that, as the third example – the Nuremberg Files – shows, a Version 1.0 approach is not necessarily the domain of progressive or left-wing forces. Lastly, what I also want to note here is how much people rely on *old ideas* in using the *new media* – how much Internet activism is about backing into the future.

LabourStart

If there's a killer app for Internet politics, then it's email. Consider the kinds of letter-writing campaigns and rapid response tactics made famous by Amnesty International, and how much more readily and easily people can take part online. One example of effective use of email alerts was generated by Eric Lee, an expat American based in the UK, from where he runs LabourStart – a news and resources website for trade unionists.[19]

Christmas 1998: Eric Lee sees a story on the BBC website about a Chinese union organiser who faces a treason trial and a likely death penalty for organising protests by workers who'd lost their jobs. The trial is scheduled to take place on Christmas Day, a day which Lee suspects has been chosen to avoid international attention. Lee tries to find out more information, but can't find details on any of the obvious sites, such as Amnesty, Human Rights Watch, or other union groups. So he sends out a bulk email to his LabourStart list, which has close to 1000 names on it. Here is part of his message:

> I'm sorry for disrupting your Christmas holiday weekend, but news came in yesterday that demands the immediate attention of trade unionists everywhere: the Chinese government is standing labour activist Zhang Shanguang on trial beginning tomorrow (Sunday). His 'crime' consists of organizing laid-off workers. If convicted, Zhang faces the death sentence. In China, as you know, death sentences are carried out regularly. Trials are swift. There may be very little time to act... Please make sure that your union and political figures close to the labour movement speak out and mobilize quickly. Please inform me of what you have done – I will try to link to every statement issued by trade unions and others in support of Zhang Shanguang in the coming hours and days. If anyone has concrete suggestions on how to attempt to persuade the Chinese government to hold back from its threat to murder this courageous man – please send these on and I will post them to the LabourStart website...

'The reaction was swift,' says Eric Lee. 'Within minutes I had the first message. A group in South Africa sent a protest message. Then a group of activists in Sheffield contacted their local Member of the European Parliament. Then the co-ordinator of Austrian LabourNet translated my message into German and sent it out to trade union contacts, as well as sending a faxed protest to the Chinese embassy in Vienna. An official of a Swedish financial workers union wrote that he'd sent it on to the daily newspaper of the labour movement there and to all key officials of his union. The secretary of the New

Democratic Youth of Canada sent out an urgent message to a long list of contacts, including MPs. And so on. Within hours, Chinese embassies in half a dozen countries had received protest messages. And none of the official organisations of the labour or human rights movement had yet done a thing – everyone was away on their Christmas break.'

The trial went ahead and the Chinese unionist was sentenced to ten years' jail. 'Not fun,' as Lee says, 'but better than death.' As the holidays wound up and people went back to work the issue generated more and more protests, with some national unions – Canada's post office workers, for instance – making formal protests to China. 'What we had done,' says Lee, 'was a global mini-mobilisation that succeeded in spreading the word and possibly – I emphasise possibly – contributed towards saving the life of an imprisoned trade union activist. And we did it using the web and email.'

While Lee credits the web here, its use in this example was mainly confined to alerting him to the case of the Chinese unionist, with the real action taking place through email. But the web is also an important tool for Net activists, and one which we'll look at throughout this book. When we talk about 'the Internet' we tend to gloss over the points of difference between applications. In terms of their uses, websites, newsgroups and chat, to take just three examples, have as many points of difference as they have points of similarity. Most of the early claims about the Net's political potential were in fact claims about email and other text-based applications such as Usenet. They pre-date the web's ascendancy to its current central position in most people's experience of cyberspace. But the web too can offer a great deal to campaigns – especially when combined with email.

A Chinese CNN?

May 1998: In the chaos surrounding the ouster of Indonesian President Soeharto, ethnic Chinese bear the brunt of much violence. Chinese trying to leave Jakarta have their cars set on fire, while rampaging mobs scour Medan shouting 'Destroy the Chinese'. Even Soeharto crony

Liem Sioe Long, the Chinese head of the country's biggest conglomerate, is forced to flee the country, as his home is looted and burned. Jakarta's airport begins to resemble a refugee camp, and ethnic Chinese besiege the Australian embassy, hoping for visas. Allegations of an organised campaign of mass rapes of ethnic Chinese Indonesians begin to circulate.[20] *Among the unpredictable consequences of this violence is the organisation and co-ordination of worldwide protests by a group of overseas Chinese calling themselves the World Huaren Federation (hereafter 'Huaren').*[21]

Huaren is a non-profit organisation that describes itself variously as a 'Chinese CNN' and as a 'webzine with attitude'.[22] The main focus of the Huaren website is their campaign to raise awareness of anti-Chinese discrimination, a campaign which fully exploits the Internet's capacity for forming horizontal linkages. Founded online by Joe Tan and Dan Tse, who live in New Zealand and Canada respectively, the group lists four main objectives as its 'mission statements': to provide a forum for overseas Chinese; to develop the self-image and dignity of Chinese people; to study the impacts of specific political contexts on the Chinese diaspora; and to work with other organisations focused on improving race relations in multicultural countries, specifically in South East Asia.[23] Some twelve people are currently involved in organising and co-ordinating activities; beyond that, the group relies on supporters to help with promoting the site and circulating messages.

'Huaren.org was probably the first active group to use the Internet technology to specifically broadcast the concerns in Indonesia,' says Dan Tse. 'Other people were then encouraged to add their effort through news groups and discussion groups.' Launched in February 1998, the site itself pre-dates the actual riots; aware of the deteriorating Indonesian situation, the group's founders began with an email campaign in January.

'We decided to start the campaign by emailing,' explains Joe Tan, 'to alert Chinese communities of the atrocity in Indonesia. We asked them to fax or email their protest to all Indonesian embassies, to other UN bodies, Chinese politicians in China, Taiwan and Hong Kong. As more Chinese learned of the disgusting atrocity and racist discriminations, they spread the news to friends in every part of the world.'

The campaign was stepped up following the first major anti-Chinese riots and reported rapes, on 13 May. Huaren received email inquiries and contacts from throughout Asia, Europe and North America, and Chinese communities in Western countries began to form local chapters. More importantly, they also began to act on what they'd learned online.

26 July 1998: In Hong Kong 2000 people gather outside the Indonesian consulate to protest against the ethnic violence, before moving on to a candlelit vigil. Two days later, hundreds gather outside a hotel in Manila, where Indonesian Foreign Minister Ali Alatas is attending a meeting. On 7 August, co-ordinated demonstrations take place in Washington DC, Los Angeles, Chicago, Houston, San Francisco, Toronto and Vancouver, as well as New York, where the crowd of 7000 is claimed to be the largest-ever turnout of the city's Chinese community. The next day, the protests reach Helsinki, Auckland and Tokyo. Demonstrations continue to spread over the next three weeks, including a gathering outside the Indonesian embassy in Beijing, which is one of the largest student actions in the city since June 1989. The Beijing students brandish photographs of alleged rape victims downloaded from the Internet.[24]

Such photographs are a key feature of the Huaren site, where some appalling pictures of alleged victims are organised under headings such as 'Torched Bodies', 'Raped and Burnt to Death' and 'More Ugly Scenes'.[25] Accounts from survivors and witnesses are archived alongside media coverage, including lengthy articles from *Asia Week* and the *New York Times*. Banners of support are reproduced for supporters to place on their own websites, and a chain letter email form is provided, along with a Yellow Ribbon campaign for other sites to copy and display. The central link on the main page leads to a summary of demonstrations and protests around the world that have been sparked by the Net campaign in which Huaren played a key role.[26] The sources are a mix of the established media (Reuters, the *South China Morning Post*, Associated Press, Dow Jones) and named contributors or Huaren editorial staff.

This mix of original and established sources is part of the group's complex attitude towards the media. On the one hand, the group's whole existence is based on their claim that news media neglected these Indonesian issues: 'The Western news media,' says Joe Tan, 'paid little attention to the blatant atrocities in broad daylight and such bias is annoying most of us.' But on the other hand, Huaren also rely heavily on the established media's coverage of those same issues to add authority to their own claims. Their large 'What's New' section archives news articles from media sources around the world on an ongoing daily basis; elsewhere in the site, contributions from supporters (victims' accounts and reports on protests in, for example, Sydney) are mixed with reports from the BBC and CNN, and from newspapers around the Asia-Pacific region and the US.[27]

Huaren's relationship with the media is a pragmatic symbiosis. While they criticise the established media, they also depend heavily upon them for sources, and for the authority conferred by the association of established international news organisations, interwoven with their own in-house content and supporters' contributions. It's possible that this is simply naive and unconsciously contradictory. It's equally possible, however, given the overall sophistication of the group's communications, that it's a shrewd and conscious strategy, aimed at simultaneously utilising their visitors' perceived mistrust of the mediascape while at the same time drawing from it what legitimacy they can.

The significance of Huaren's media strategy can be illustrated by a comparison with the Tiananmen Square student movement, itself widely celebrated in the Western media for its use of new communications technologies. Academics Constance Penley and Andrew Ross have observed the irony of this, noting that much of the communication in the Tiananmen movement was one-way, as the students struggled to collect what information they could from Western sources. The students' use of fax machines, for instance, to bypass government controls and receive news from overseas, has become well-known.[28] But a consequence of this reliance on Western media was that 'very little information about the students' actual demands, desires, and strategies was directly relayed', write Penley and Ross. 'The students were receivers of Western information – more often than not, Western definitions of their own aims – rather than broadcasters of their own messages.'[29] Huaren, by contrast, use Western media sources where they can, but also mix in their own content, creating a more complex output than that achieved by the Chinese students in 1989. This illustrates some of the complex outcomes that Net activists can achieve through informed use of the technology.

More complex still are the potential impacts of such Net use at the level of the nation-state – particularly states which attempt tight control over information flows. In this respect, Huaren illustrate what anthropologist Arjun Appadurai calls the 'diasporic public sphere', a cultural space which develops out of mass migration and mass mediation:

> As Turkish guest workers in Germany watch Turkish films in their German flats, as Koreans in Philadelphia watch the 1988 Olympics in Seoul through satellite feeds from Korea, and as Pakistani cabdrivers in Chicago listen to cassettes of sermons recorded in mosques in Pakistan or Iran, we see moving images meet deterritorialized viewers. These create diasporic public spheres, phenomena that confound theories that depend on the continued salience of the nation-state as the key arbiter of important social changes.[30]

Such diasporic public spheres are increasing in number and importance as new producers and audiences link across national boundaries, creating new audiences and producers, new publics. State efforts to control media flows will continue to be undermined by the kinds of global flows visible in the Huaren campaign: in Saudi Arabia, Internet cafés bypass the government's attempts to limit the range of sites available by connecting to the Net through direct satellite links.[31] In Bangladesh, the 1994 international support campaign for writer Taslima Nasreen when she was charged with blasphemy used the Net to exchange information and co-ordinate representations to the government; email connected Bangladeshi human rights activists to PEN and Amnesty, while the Bangladesh.Soc.Culture newsgroup provided a vehicle for sharing information across otherwise tightly controlled borders.[32] In Malaysia, online newspapers such as Malaysiakini.com and HarakahDaily.com challenge the Mahathir government's tight controls on media independence, while dissident opinions find outlets on sites such as FreeMalaysia.com; the irony here is that the Mahathir government promised not to censor the Net as part of its drive to recruit US technology companies for the country's Multimedia Super Corridor project.[33] And as we'll see in chapter 3, the Serbian government's attempts to silence independent radio station B92 backfired when the station moved to the Net.

Did the Huaren campaign produce results? In November 1998, the Indonesian government's official inquiry confirmed 66 rapes, the victims of which were mainly ethnic Chinese women. The commission found that not all of the more than 400 reported rapes could be confirmed. While its report did not reach a definite conclusion as to whether or not the rapes had been organised, the implication remains – one commission member drew attention to findings that the military had been involved in instigating and inciting rioting. The report called on the Indonesian government to sign an international declaration against racial discrimination.[34] Huaren argue that this official inquiry would not have even been held without the pressure of the global demonstrations organised through their site and others like it – there is no way, of course, to confirm or deny this.

One interesting aspect of the evolution of Huaren's campaign is that Joe Tan and Dan Tse felt that the newsgroup discussions of the Indonesian situation in January weren't enough. The information had to be moved to the web to have a chance of reaching a wider audience – newsgroups didn't offer a sufficiently public forum. In their aspiration to become 'a Chinese CNN', the group felt the need for a higher profile and a co-ordinated centre – the website offered a more public orientation than the highly structured, semi-private environment of news groups.[35] It's easy to overstate the extent to which the web is accessible to a universal public. Without incoming links and search engine registration, any website can go unvisited. But this public potential of the web has itself been used by more than one campaign in another way – as a *threat*.

The Nuremberg Files

23 October 1998: Obstetrician Barnett Slepian is shot dead through the window of his kitchen in Buffalo, New York, by a sniper with a high-powered rifle. Within a few hours, his name has been crossed off a list of practising abortionists maintained on a website called the Nuremberg Files – an anti-abortion site which encourages supporters to show how pro-life they are by shooting doctors.[36] *The main feature of the site is a list of the names of more than 250 doctors. A helpful key explains that names in a black font are 'working', grey names are 'wounded', and names with a strikethrough signify a 'fatality'.*

Visitors to the site at the time of Slepian's death found that he was fatality number four. They also found invitations to email more details on individual doctors to add to the site's files. Detailed dossiers on some doctors had been compiled – one included its subject's driver's licence number and her children's names, dates of birth and schools. There were form letters to send, including one which asked practising abortion doctors to send in a favourite photo of themselves for inclusion on the site. More recently, the site has launched a webcam project to film 'people going in or out of baby butcher shops in your city or town'.

Did the campaign work? Only too well. In the Nuremberg Files case, the consequences for these doctors of being identified by the site were very real and very serious. In February 1999, a federal jury in Oregon awarded more than US$100 million against the group behind the site – who include a church minister with convictions for setting fire to abortion clinics – after a number of the doctors named on the site testified that they had taken to wearing bulletproof vests, wigs and disguises since being identified. This is a testament to the power, if not the ethics, of the tactic of *outing.*

Outing's potential consequences have always made it a divisive tactic. It was pioneered by the American groups ACT UP and Queer Nation, and imported to the UK in 1991 by a group called FROCS (Faggots Rooting Out Closeted Sexuality). FROCS used it as a *simulation:* they let it be known that they had compiled a list of MPs and other public figures whom they were planning to out. But they then held a media conference at which they accused the press of hypocrisy – after running stories condemning the tactic, journalists had still turned up in the hope that names would be named. OutRage! picked up the tactic in 1995, but this backfired when one activist offered hints to an Irish journalist: the *Belfast Telegraph* then identified and contacted one hardline anti-gay MP, who promptly, perhaps coincidentally, dropped dead of a heart attack.[37]

After the 1999 court decision, the Nuremberg Files site was removed by the Internet service provider (ISP) which had been hosting it, but a number of free speech advocates around the world were to put up complete mirror versions.[38] In March 2001, a federal appeals court in San Francisco ruled in favour of the Nuremberg Files, citing the First Amendment in its judgement that the site couldn't be held responsible if it 'merely encouraged unrelated terrorists'.[39] Site operator Neal Horsley is back in business.

The Nuremberg Files is an example of a site which exploits the perceived technical properties of the Net – its public openness, its global reach, its resistance to censorship. Remembering that the development of the early Internet infrastructure was funded by the US military, it's an irony that here's an example of people using the medium as a *weapon.* There are other examples of similar campaigns: in the US, there have been some cases of unions posting photos of scabs on their sites,[40] while in Australia the gun enthusiasts at Lock, Stock & Barrel are building an online gallery of pictures of politicians' houses, along with their addresses.[41] Of course what often happens is that this kind of site gets closed down. But in the meantime, they're very effective in attracting *media attention* to a cause. Even *hostile* coverage and forced closure mean that the site and its agenda are publicised, and this is a lesson that other groups are almost certainly going to pick up on.

Backing into the future

In terms of meeting their objectives or drawing media attention to their cause, all of these examples worked. But one striking thing about them is how much they rely on the tactics which had proven most effective in a pre-Net world; as we'll see in later chapters, this is a comment which also applies to many other Net campaigns. The tactics used in those three campaigns, and in the Seattle events I described earlier, are not the only ones to be updates of pre-Internet tactics. It's worth acknowledging, of course, that many online campaigns are continuations of existing struggles for justice and so inevitably continue to use the approaches honed offline. But it's also worth acknowledging that, so far, there's little evidence of entirely new tactics developed specifically to exploit the unique properties of the Net. This may be less a failing of online politics than an inevitable consequence of the ways in which we adapt to technological change. As Marshall McLuhan noted more than 30 years ago:

> When faced with a totally new situation, we tend always to attach ourselves to the objects, to the flavor of the most recent past. We look at the present through a rear-view mirror. We march backwards into the future.[42]

So how much of Internet politics is about *backing into the future*? And how much critical writing about the Internet is also about backing into the future?

If effecting social change of some kind is the objective, then the *strategies* will involve networking, publicising, educating, organising and mobilising. As for the *tactics* used to achieve these strategies, pretty much every approach to raising awareness of issues has now been taken online. The whole repertoire of tactics developed throughout the twentieth century, from the Suffragettes to Civil Rights, from Greenpeace to ACT UP, from Gandhi to Greenham Common, have found their digital analogues, as social activism moves into cyberspace. Letter-writing, phone and fax trees, petitions. Newsletters, newspapers, samizdat publishing, pirate radio, guerilla TV. Ribbons and badges, posters, stickers, graffiti. Demonstrations, boycotts, sit-ins, strikes, blockades. Sabotage, monkeywrenching, outing. Even online benefit gigs and virtual hunger strikes.[43]

In some cases, adapting old tactics to new media can be an inspired idea. Email, for instance, has certain properties that make it ideal for the kind of rapid response alert used by Eric Lee. These include its speed and immediacy, its ease of use and economies of scale, and the ways it can be forwarded and cross-posted across borders and time zones. But these same properties make email a poor choice for petitions, another pre-Net tactic which has been taken up enthusiastically by many activists: decision-makers are likely to be unimpressed by a haphazard list of names that arrives piecemeal, with repeated signatures or pseudonyms from people well outside their jurisdiction. Potential signatories can also be irritated and alienated by cross-posting – an email petition doesn't become any more relevant to me just because I've received it three times. But the email petition proliferates because it's an easy idea to come to grips with, because it's *familiar*.

Such familiarity, in some cases, offers certain advantages. In chapter 6, for example, we'll look at the tactic of the *virtual sit-in*. On one level, as an electronic update of an established protest gesture, the virtual sit-in is about backing into the future. But it also illustrates one of the advantages of this: if it's a familiar tactic, people can quickly see where you're coming from.

'At first movement,' says Ricardo Dominguez, a key figure in the development of the virtual sit-in, 'I think it's much easier for people to manifest themselves if they can consider that it's somehow bound to a history that they know. If I said "virtual sit-in", it had a kind of pedagogical usefulness that the term that I would prefer – "electronic civil disobedience" – did not have. And it's something that they're not afraid of – they understand what a sit-in is, they understand that civil disobedience is about non-violence. It becomes a way in which you can re-embody in cyberspace a certain lived condition which people understand traditionally from Gandhi, from Civil Rights.'

But not all instances of backing into the future are so well considered. The multimedia potential of the web, for example, tempts many into streaming webcasts of real-time video and audio. A natural choice, in some ways, in that streaming is what we can see in the rear-view mirror – TV and radio. Yet the archival nature of the web and its status as a medium which we *consult* – rather than have on all the time in the corner – undermines the effectiveness of streaming. An archive of well-chosen clips, which we can consult without conforming to the schedule of a live stream, is a more powerful use of the medium.[44] So the obvious pitfall in reaching for a familiar tactic is that the means chosen need to match the desired end. As community organiser and tactical theorist Saul Alinsky argued, Gandhi's non-violent resistance tactics were well suited to a campaign against a British administration operating within a liberal tradition – they wouldn't have worked against the Nazis.[45]

One other thing to note here is how much all the examples above rely on the mainstream media *as well as* the Internet. The Seattle protests were only a success insofar as they put the question of globalisation squarely on the wider media agenda. Eric Lee learned about his Chinese unionist from the BBC. Huaren relied heavily on established news sources. Online outing campaigns such as the Nuremberg Files have got most of their attention through stories about them in the newspapers. Internet activism is largely about raising awareness of the issues concerned, and this means getting more coverage than the purely online.

Coming up

In chapter 2 we'll build on the Versions 1.0 and 2.0 model, as we look at the idea of *interactivity* and its role in popularising the early ideas of cyberdemocracy. Interactivity is usually sold to us as something new and sexy, and somehow unique to digital media. But what 'interactive' might actually mean is usually treated as a given. So in this chapter we'll consider different dimensions and definitions of interactivity as a way of coming to grips with cyberhype. We'll also look at how traditional representative organisations – political parties – are using the Net. Readers who can't bear the thought of reading even one word about political parties, and are impatient to get back to the activist case studies, are encouraged to skip ahead to chapter 3.

CHAPTER 2

Remote Control

In February 1997 it was reported that Sony were about to introduce a new TV remote control that could itself be operated by remote control. This would replace older models of remote that had to be operated by hand. The report suggested that as it would no longer be necessary to actually reach towards the coffee table to pick up the remote, the product would mean 'greater convenience for TV viewers'. This would also enable viewers to become 'more immobile, more stationary and more physically inert than ever before'.[1]

The story was, of course, a parody from the satirical newspaper *The Onion*. But what's striking about it is how well it captures the discourses of interactivity as they routinely appear in the popular media – loosely defined, loosely deployed, the concept of interactivity implies greater autonomy and agency. But does it offer anything more than ever-smarter remote controls? 'Interactive', points out musician and digital artist Brian Eno, 'makes you imagine people sitting with their hands on controls, some kind of gamelike thing.'[2] Interactive media, we're told, will mean we can choose between different camera angles while we're watching the match, call up that new movie without going to the video shop, or click on our favourite actor's shirt when we're watching TV and order one in our size. Interactivity means never having to get off the couch.

This remote control angle is a very different vision of interactivity from the one which was central to much of the hype about Version 1.0 of the Internet. We heard how the Net would transform politics, revitalising civic debate, political participation and public life. The only problem with contemporary politics, apparently, is that they aren't *remote* enough.

So in examining such claims, and also actual political uses of the Internet, unpacking and untangling the loose uses of the much-abused word 'interactivity' is a good place to start.[3] We'll begin by considering different aspects of interactivity, then use these to examine some of the early claims made for the democratic potential of the Net. We'll then explore how this potential is being exploited – or ignored – by political parties. As central representative institutions, parties have traditionally been early adopters of communications technologies, so their Internet use deserves close examination. We'll conclude by looking at two projects which illustrate key phenomena associated with the relationships between politics and new media: first, the arguments for a form of 'direct democracy' that is facilitated by so-called interactive technologies; and second, the rise of maverick political movements such as Pauline Hanson's One Nation.

So what does 'interactivity' actually mean? The popular usages take us some of the way, but not far enough: first, the term is often used to contrast with established media – TV, radio and print are all essentially top-down communications technologies. They can get to us, but we can't easily get to them. The communication flows are mainly one-way, and while we do some of the work in figuring out what messages mean and what to do with them, we don't get to send many messages of our own – letters to the editor, talkback radio and TV phone-ins are obviously limited in scope. Early Net technologies – email, bulletin boards, newsgroups – produced so much optimism and excitement at least partly because they were seen to enable participation on a scale, and to a degree, not possible with established media. Any of us could, in theory, take part. An interactive technology was one in which we could get to them as easily as they could get to us.

But does this apply to the world wide web? Not necessarily. Many websites are no more interactive than a jukebox – the selections are predetermined and we simply click on our choice. They can get to us, but we can't readily get to them. And since the early excitement about the Net and interactivity, the *web* has become central to most people's online experience. But for a website to be interactive in any meaningful sense, it has to be designed with two-way input as a goal. Many aren't. Instead, many websites are designed to preserve the one-sided advantages of the broadcast model, with promotion, persuasion and propaganda as the goals. But even using words such as 'participation' and 'broadcast model', as I just did, doesn't do enough to clarify the concept. In practice, there are different degrees of interactivity, not all of which relate to the Internet as a Version 1.0 open system – most, in fact, are Version 2.0.

Every Friday, for example, I receive an email from *The Economist* magazine, telling me which stories from the latest issue are available on their website, and offering brief summaries. I can choose whether or not to follow these up, and can do so in my own time. If I do get around to visiting the site, I can select from a menu of stories and might end up reading something I hadn't planned to, before drifting off to a different site – behaviour which will be logged by *The Economist*'s registration system. If I think a friend might be interested in what I've read, I can forward the link and later discuss the story in an email exchange which descends into vitriolic flaming, leading to a meeting in the pub that night to repair the damage over a few beers.

Each step of this process could be called interactive, but the nature of the interactivity is different in each case. After subscribing to the magazine's email update, I have no control over when the digest is sent to me, or over its content. Choosing to subscribe to the service is an example of what communications theorist Jens Jensen calls *transmissional* interactivity – it allows me a degree of control, or rather *choice*, over what is otherwise a chronologically fixed or programmed information flow. This is the closest thing to a broadcast take on interactivity.[4]

When the site monitors and logs my movements between stories, again I have no say over this beyond setting my browser to reject cookies, a decision which might restrict my access within the site.

Jensen labels this *registrational* interactivity. It's about collecting information on visitors – we produce this information, but have little or no say in how it is used and distributed.

Once I get to the site, I can choose between articles but I have no say over their content and no real provision for input – I could write to the editor, if I could be bothered, but I couldn't modify the story in any way. The choices are predetermined. My interactivity here is primarily *consultational* – I select, I choose, I use *The Economist* jukebox.

Only in the last phase, the email discussion with a friend, do I have a potentially equal relationship, and the ability to influence and contribute to the *content* of the exchange. This last phase is an example of what Jensen terms *conversational* interactivity. Only with conversational interactivity do we get to the idea of a two-way communication flow, with both partners producing and inputting their own information; and, often more than this, working to *create* something. This was the pattern of interactivity which generated much of the early excitement about the Internet and democracy. Not only is it the pattern which is most useful to a Version 1.0 position, but it also highlights the extent to which the other forms of interactivity are really part of Version 2.0: each relates to a closed system, to a vision of Internet use as consumerism.

Conversational interactivity should be thought of as separate from the other forms. As we'll see below in the section on One Nation, it does have its limitations. But these can be worked through as part of the process of creating new spaces for debate and action (and perhaps, eventually, new *forms* of debate and action). It may be, in fact, that we shouldn't think of this conversational dimension as 'interactivity' at all – the term is so thoroughly in hock to a Version 2.0 world view. Eno argues that 'interactive' is the wrong word:

> The right word is 'unfinished'. Think of cultural products, or art works, or the people who use them even, as being unfinished. Permanently unfinished. We come from a cultural heritage that says things have a 'nature', and that this nature is fixed and describable. We find more

> and more that this idea is insupportable – the 'nature' of something is not by any means singular, and depends on where and when you find it, and what you want it for. The functional identity of things is a product of our interaction with them. And our own identities are products of our interaction with everything else.[5]

The strength of Eno's idea of the 'unfinished' is that it opens up a different space for us to think through our relationships with new media. If the 'interactive' is about *consuming* media in (more or less) novel ways, the 'unfinished' is about people *making* new media for themselves.

This point – that conversational interaction is not an end in itself, but a means to a creative end – is one that we can see in some of the ideas of Tim Berners-Lee, creator of the world wide web. He defines 'interactivity' as not just being about choice, but about *creativity*:

> We ought to be able not only to find any kind of document on the Web, but also to create any kind of document, easily. We should be able not only to follow links, but to create them between all sorts of media. We should be able not only to interact with other people, but to create with other people. *Intercreativity* is the process of making things or solving problems together. If *interactivity* is not just sitting there passively in front of a display screen, then *intercreativity* is not just sitting there in front of something 'interactive' [emphasis in original].[6]

This idea of *intercreativity* is key in analysing Net activist campaigns, as we'll see in subsequent chapters. The projects we'll look at are all, on one level, about people not only interacting, but creating together: most often, about creating an alternative to the established media – an alternative that is an open, Version 1.0 space. So, on the one hand, we have *interactivity* as an extension of the established media – not a new concept, so much as a new spin on an old one; a hot new word to use in ads. And, on the other hand, we have the *conversational*, the *unfinished*, the *intercreative* – the possibilities of a future that, in this case, we *don't* back into; one that may not look like what's gone before. Glimpses of what a Version 1.0 Internet built on such possibilities might look like can be found in the new media networks built out of Radio B92 (chapter 3) and in the Indymedia movement (chapter 4). Each of these illustrates intercreativity, and each is a space of ongoing possibility.

Cyberhype

The early enthusiasm about the democratic potential of the Internet centred around the conversational dimension. Access to information would create and sustain virtual community, but that community would be built and developed through discussion and debate.[7] And, as we'll see below, the successes of such early intercreative communities would be central to much of the writing about the early Net.[8] In this version there would be a resurgence of community flowing from a new kind of individual empowerment – cyberspace would be the digital *agora,* thanks to a technology which was decentralised, egalitarian and non-hierarchical; a technology which enabled us to be judged by our contributions – our *words* – rather than by meatspace variables such as age, gender or ethnicity. It's in the emphasis on our words that we see the importance of the conversational dimension.[9]

Major contributions to this vision came from John Perry Barlow and Mitch Kapor, co-founders of cyber-rights lobby group the Electronic Frontier Foundation. Barlow, who takes the credit for applying novelist William Gibson's term *cyberspace* to actually existing computer networks,[10] was moved to describe computer-mediated communication as 'the most transforming technological event since the capture of fire'.[11] He claimed, on behalf of those on the Net, that governments 'have no sovereignty where we gather'.[12] In 1993 Kapor wrote that life in cyberspace 'seems to be shaping up exactly like Thomas Jefferson would have wanted: founded on the primacy of individual liberty and a commitment to pluralism, diversity, and community'.[13] But the 1990s, warned Kapor, would be crucial to the democratic potential of cyberspace, as it confronted the realities of regulators and commercial forces. By the end of that decade, his optimism had been somewhat diminished.

'My original views were hopeful and, with 20–20 hindsight, naive,' Kapor told me in 2000. 'I placed too much faith in the power of technology itself and did not give enough credit, positive and negative, to the many ways in which we, as a society, have merely imported our enthusiasms and prejudices from *terra firma* to cyberspace. The increasing commercialisation of the Internet has certainly played a major role in giving it its character, but I do not see it as a cause as much as an effect of who we are and how we conduct ourselves.'

Probably the best-known proponent of cyberdemocracy is Howard Rheingold, whose book *The Virtual Community* is partly an account of his experiences in the text-based conferencing system of the WELL (Whole Earth 'Lectronic Link).[14] Rheingold offers a wealth of evidence to support his thesis that 'whenever CMC [computer mediated communication] technology becomes available to people anywhere, they inevitably build virtual communities with it'.[15] Such grassroots community organising, he argues, offers people enormous potential power: 'intellectual leverage, social leverage, commercial leverage, and most important, political leverage'.[16]

Many of Rheingold's examples are persuasive, though the virtual bonds of the WELLsters often appear to be reinforced through real-world meetings, parties and phone calls, which does raise doubts about the model's wider usefulness. Early WELL members' online interactions, in other words, were reinforced by old-fashioned In Real Life interaction. Like many Internet activist tactics, Rheingold's book is very much about backing into the future. Seeking to conceptualise online life, he reaches back to a mythical, prelapsarian America, an idealised community – the subtitle of the book in one edition is 'Homesteading on the Electronic Frontier'. And he hangs much of his argument on Habermas's concept of the public sphere – an idealised conceptual arena for the formation of public opinion – which in itself backs into the future by harking back to a dodgy golden age.[17] While still a firm believer in the potential of the earlier text-based Net,

Rheingold, like Kapor, is now less optimistic about the commercial, graphically based web, although he cautions against imagining that the latter has completely replaced the former.

'You can still get to newsgroups, chat rooms, web conferences, and mailing lists through the web,' Rheingold points out, 'so social discourse has not gone away. In general, the overwhelming attention and resources directed at commercial enterprises online is having the same effect that the privatisation of everything in the offline world is having: the public sphere is increasingly lost in the private sphere.'

But, crucially, developments in the technology and its uses, since *The Virtual Community,* mean that most people's experience of cyberspace is now situated around the web. And the web is not the ideal vehicle for the conversational dimension of Net use. In the ways in which it's most commonly used, it's about transmission, registration, consultation – it's about the interactive, the Version 2.0 model. The back channel for our input is often small, more often nonexistent. *The Virtual Community* is one of the two founding pillars of cyberculture studies, along with Sherry Turkle's *Life On The Screen*, which popularised concepts of identity fluidity and aliasing, of online gender-bending and self-construction. They're both good books, but their influence has been a problem, in that they both address pre-web text-based interaction. Many of those following Rheingold and Turkle applied their ideas a bit too loosely to the web – or, more often, ignored the web in favour of writing yet another article about digital dungeons-and-dragons games.

This is not to say that Rheingold was wrong. Rheingold was actually never as naively starry-eyed as he is sometimes claimed to be; his overall optimism in *The Virtual Community* is significantly tempered – he frequently qualifies his analysis with reminders that the situation he describes is a temporary one, and that the darker potential of the technology – surveillance and commercialisation, for instance – is equally real. His reputation as a utopian perhaps derives from his pragmatic optimism.[18] Asked in 1999, for instance, whether there was still scope for the many-to-many communication potential of the Net to become dominant, or whether it would inevitably be undermined by commercialisation, Rheingold answered: 'Both! Where did Yahoo! come from? Two college kids. New media come from everywhere, and some of them become economically dominant and start acting the way big money enterprises act. That doesn't stop anyone from putting up their web page and publishing, broadcasting, or discussing anything they wish.'

The seeming utopianism of writers such as Rheingold and Barlow, and the uncritical iteration of their ideas through the media, produced a whole sub-genre of writing. It's typified by *Wired* magazine, and by the computing section of a newspaper near you – a mix of product placement and speculation based on other product placement and speculation. Such writing also produced a backlash, a sub-genre of critical commentary aimed at what media theorist McKenzie Wark labels *cyberhype*.[19] In an important contribution to the critique of cyberhype, media theorist Darren Tofts argues that the pace of technological change is not outstripping our capacity to manage it, but the *hype* surrounding such change has 'achieved escape velocity, overcoming the gravitational pull of actuality'.[20]

As Tofts points out, one result of this kind of cyberhype is that the electronic ante is forever being upped. The dizzy rhetoric surrounding still-unrealised technologies is usurped by the fanfare for other unrealised technologies, so that 'the data-suit supersedes the head-mounted display and data-glove that most people have never used in the first place'.[21] The commentary on future possibilities is so ubiquitous that it's as though that future has already arrived. At one extreme, this leads to serious social scientists producing discussions of cyberspace which have less to do with the actuality of an increasingly commercialised Internet than with fantasy prototypes as featured in *Wired*.

Think back, for instance, to the claims we heard for Virtual Reality, only the day before yesterday. One writer described his vision of a town hall meeting in which wearing Virtual Reality helmets and gloves provides 'an exciting alternative for dull gatherings on the basis of flat maps'.[22] Another saw these same helmets and gloves as the key to a time in which 'real space loses any capacity to constrain social arrangements' – as a happy consequence of using 'masking devices that fit screens to the eyes, speakerphones to the ears and control "gloves" to the hands and feet'.[23]

What is being proposed in this kind of commentary is little more than a form of democracy-through-gimmickry, as though electronic bread-and-circuses would somehow get everyone interested in public affairs rather than providing mere distractions. Not that there's much new in this claim itself: technological novelty has often been seized upon in the past by those claiming to sense its democratic potential. For example, a 1973 report on cable television by the US National Science Foundation suggested that 'cable can become a medium for local action instead of a distributor of prepackaged mass-consumption programs to a passive audience'.[24] Tell that to a subscriber. Nor are such claims restricted to the promotion of new media: technology critic Langdon Winner offers a useful list of twentieth-century innovations which were hailed in their time as new forces for democracy – nuclear power, phosphate fertilisers and the factory system.[25]

But it's important to untangle the different threads of cyberhype discourse. There was firstly a cyberhype of the Net Version 1.0, typified by the writings of Rheingold and Kapor. And Version 2.0, of course, had its own strain of cyberhype, and still does, as high-profile commentators argue that the market can work its magic if governments adopt a hands-off approach.[26] This is the cyberhype of the virtual mall, in which the processes of commercialisation are hidden behind a veil of technological determinism through which changes for the better will somehow just happen, as though of their own accord.

One example is the 'Magna Carta for the Knowledge Age', produced by a group including software industry commentator Esther Dyson and futurologists Alvin Toffler and George Gilder. This manifesto argues that governments have to leave the information market free for 'dynamic competition'. Governments, these authors suggest, are essentially outdated technologies of the Industrial Age, built around standardisation and centralisation: factory-model concepts, with no place in an era where images, symbols and data flow across borders and time zones.[27] But if governments cede influence not to private citizens but to global corporations, for whom is this a good outcome? (This question is addressed by the some of the groups examined in later chapters.)

Another key contributor to Version 2.0 cyberhype is Nicholas Negroponte, director of the MIT Media Lab. Negroponte outlines a future – ten minutes away – in which 'your right and left cufflinks or earrings may communicate with each other by low-orbiting satellites'.[28] His optimism is often engaging, but it's hard to read as anything other than corporate boosterism, as the promotion of a future to be determined by the stock options. More recently, for instance, Negroponte has been predicting that household appliances and toys will soon generate more Net traffic than humans: 'Barbie dolls,' he has been quoted as saying, 'will be able to educate themselves by downloading different language software from the Internet.' Barbie's makers, Mattel, are among the 170 corporate sponsors of Negroponte's Media Lab.[29]

Cyberhype discourses are a symptom of the tensions between the Net Versions 1.0 and 2.0, between the democratic potential of an open system and the commercial potential of a closed one. The cultural critics and academics who took aim at cyberhype were primarily attacking the tendency for Version 1.0 rhetoric to be co-opted into Version 2.0 marketing plans. For example, both Vivian Sobchack and Mark Dery critiqued early cyberculture magazine *Mondo 2000* for its shameless mix of hard-sell consumerism and warmed-up 1960s idealism.[30] Dery described the political alternative offered by the magazine as 'a fantasy of escape from the constraints of time, space, and the human body rather than a reasoned, realistic response to the politics of culture and technology in peoples' everyday lives'.[31] And McKenzie Wark pointed to the irony that the 'revolution' promised by such magazines involved neither art nor politics, but consumer goods, labelling the marketing of cyberculture 'the first off-the-shelf Romantic revolution'.[32]

Version 2.0 rhetoric arrived early, with the streamlining of the Net into the *information highway*. Cultural critic Arthur Kroker has argued that this shift in metaphors was an attempt to impose some kind of linear order onto cyberspace: 'to liquidate the sprawling web of the Internet in favor of the smooth telematic vision of the digital superhighway'.[33] The linear metaphor of the highway, with its modernist connotations of nation-building through technological infrastructure, neatly co-opted the libertarian 'frontier' rhetoric of the early champions of cyberdemocracy.

Looking back at this rhetoric, the corporate colonisation of cyberspace was obvious from the beginning. Some of the most pervasive metaphors of libertarian cyberhype come directly from the heyday of Western imperialism – cyberspace was an 'electronic frontier'[34] to be won through a war fought 'in the trenches of hyperspace'.[35] Bill Gates, who ought to know, compared the Internet to a gold rush, adding that 'Fortunately, this is a gold rush where there really is gold.'[36] One newspaper headline helpfully tied several of these metaphors together, reporting that 'The cyber-sheriffs want to tame the lawless frontiers of an electronic gold rush'.[37] Technology critic Ziauddin Sardar dissected these metaphors, reminding us of what happens to frontiers: 'Once a new territory has been colonised, it is handed over to business interests to loot'.[38]

Each of these critiques of cyberhype was an important contribution. But it's worth noting that what they were criticising was the use of Version 1.0 rhetoric and key words to sell hardware, move software, shift units. In the marketing frenzy to pitch cyber-everything, the sexy notions of democracy and empowerment were sampled and looped, then cut-and-pasted into a commercial remix. But their original use in Version 1.0 writings shouldn't be dismissed too quickly. There was certainly a strong element of technological determinism to Version 1.0 cyberhype, but the enthusiasm of writers like Rheingold, Barlow and Kapor needs to be set back in its pre-web context.

When Rheingold published *The Virtual Community*, for example, Internet communications were largely text-based and small-scale.

The big media corporations were only beginning to pick up on the potential of this stuff – as indeed was Microsoft – and if such writers believed that cyberspace offered democratic potential, then that was because the cyberspace they were describing had itself been developed through a kind of grassroots activist process. Rheingold described this process as 'a kind of speeded-up social evolution'.[39] He was talking about the development of the WELL, but the description can be extended to the wider Net of the time – and to the grassroots, bottom-up negotiation and evolution of what the norms and standards were to be online; of how this community should be related to wider society; and of how it should be regulated, if at all.

This was the cyberspace of free software, of the hacker ethic that 'information wants to be free'. It was the cyberspace that developed netiquette. It was the cyberspace of identity fluidity (and deception), which contributed digital folktales such as the infamous 'rape' in LambdaMOO,[40] or of the male psychiatrist 'Alex' establishing relationships with women online in his disguise as the wheelchair-bound 'Joan'.[41] What's interesting about all these well-known examples is that they point to a grassroots process of decision-making, boundary-drawing, cultural evolution – a kind of activism, though one which was concerned less with *changing* a society than with *creating* one: a society of ongoing possibility – one whose strength was that it was, in Eno's sense, unfinished.

When this early Net culture began to shift its emphasis from its own internal politics to using the Net to effect social change in the offline world, many of the culture's major early activist campaigns were *about* the Net itself: about privacy, regulation, free speech and access. These were campaigns aimed at protecting and fostering the openness of the Net – campaigns which saw the darker potential of Version 2.0 coming.

A clear example is the formation in June 1990 of the Electronic Frontier Foundation, a non-profit lobby group concerned with digital privacy and free speech issues, created partly in response to US federal authorities' sweeping arrests of hackers and crackers.[42] Other significant mobilisations included the controversy over the 1994 mass spamming of Usenet newsgroups by the immigration lawyers Laurence Canter and Martha Siegel;[43] the volunteer effort of NetDay 96, which aimed to connect every school in California to the Net;[44] and the mass blackout of hundreds, possibly thousands, of websites in February 1996 in protest at the signing of the US Telecommunications Reform Bill (Communications Decency Act) which, designed to 'protect children' from digital porn, struck many as an unconstitutional attack on free speech.[45] This blackout (an online strike) and the related development of the first online ribbon campaign (the blue ribbon symbolising free speech) were also among the first instances of rear-view mirror tactics, of activists applying old ideas to the new media environment.

Another key event was the successful Net-based campaign which halted the production of MarketPlace, a database which would have harvested consumer details for sale to commercial interests. Proposed by Lotus – the company founded by Mitch Kapor – MarketPlace was scrapped after a heated online campaign involving widespread privacy debates, electronic petitions and form letters, and some 30,000 people approaching Lotus to demand that their personal details be excluded.[46] The privacy implications of this commercial product generated enough bad publicity for it to be withdrawn. But while the withdrawal was hailed by some as a 'victory for computer populism',[47] it would be a mistake to attribute the success of this campaign solely to the Net; or, rather, it would be a mistake to discount the role played by the mainstream media in contributing to the pressure on Lotus. The Net campaign was sparked by an article in the *Wall Street Journal* which highlighted the privacy issues,[48] and established lobby groups played an important part in mobilising opposition and spreading awareness. What I want to highlight here is that, from the beginning, Net activism has never been *exclusively* Net-centred. The MarketPlace campaign made spectacular use of the Net to increase awareness and co-ordinate tactics, but this was only one part of the wider media event. Such Net activism should always be considered as an aspect of broader media activism: as a means of spreading wider awareness of an issue and mobilising public support and pressure, not as an end in itself.

The above campaigns were about protecting the digital commons. More than this, they illustrate the difference between the unfinished Version 1.0 and the interactive Version 2.0. The MarketPlace example, for instance, was on one level a campaign *against* an aspect of interactivity – its registrational dimension. It was a campaign about preserving the open space of the nascent Net. When people started to flex their virtual muscles in the political arena, the issues which caught their imagination were those which directly affected the early Net culture they'd been engaged in *creating*.

All tomorrow's parties

So if the conversational dimension was the pattern most significant to early claims about the political potential of the Net, we should consider to what extent this pattern is facilitated in Net activist projects. One logical starting point for such an examination is the use of the Net by established political parties. As traditional representative institutions with a central role in public life, and every reason to adopt new communications technologies, how have they responded to the potential of the Net?

Visiting the website of a typical major party is like watching a television tuned to a dead channel. With a heavy emphasis on fundraising, appeals for money are the one area in which these sites excel. For a while the website of Labour in Britain offered more details to potential advertisers than to potential members; their campaign site for the 2001 UK election offered an admirably economical navigation menu: 'Join Us. Volunteer. Donate. Shop.' Every time you clicked on a new page at Al Gore's presidential campaign site, a new window asking for donations appeared. Political participation becomes reduced to making donations, and active citizenship is recast as active consumption. For a time the main menu of the US Republican National Committee site included a button marked 'online activism' – clicking it took you straight to the gift shop.

Survey the sites of Australian, British or US parties and you'll find Version 2.0 interactivity in action.[49] If Net activists are driving into the future while looking in the rear-view mirror, party strategists are busy painting over the windscreen. The transmissional model of interactivity is dominant – we can read speeches, sift through media releases, watch video clips and ads, play games, and of course contribute money. But we can contribute little else. The conversational dimension is close to nonexistent. Instead, there's a conscious effort to apply the tried-and-tested models of broadcast politics and information control, of image and spin, to the new medium. This involves all the forms of interactivity, but not the feature which is most important to a Version 1.0 Internet. Political parties online are, in fact, very much part of Version 2.0 – their tight management of information flow and participation is a micro-level example of the web as a closed system.

Two main areas illustrate this. First, we'll consider how parties make it difficult for us to contact them, and next to impossible for us to connect with each other. And second, we'll look at how they apply the broadcast paradigm, not just in terms of one-way communication, but also by offering content which approximates entertainment forms.

At the time of writing, no major Australian or British political party website offered any kind of open public discussion facility. There are no chat-rooms, no bulletin boards. Not only does this restrict the potential for citizens to communicate their views to the party; it also means that those interested in networking with others who share their political tastes have to look elsewhere to make contact. While the party sites are to greater or lesser degrees interactive, they don't allow for the conversational dimension. In Australia, both the Liberals and the Australian Labor Party (ALP), for instance, offer a subscription service for email updates – this is interactive, in the transmissional and registrational senses. The IT consultant who designed the ALP site described it like this: 'The Internet is not like television, where you sit back and get the action thrust at you. It's like an intimate conversation on the phone.'[50] A nice line, but in practice this appears to be just about *style* – the presentation will be informal and friendly, but an 'intimate TV commercial' might be nearer the mark.

Some US sites, such as that of Al Gore's presidential election campaign, make more concessions to the conversational dimension, offering a limited forum. But even this is still closer to the jukebox model than to a real debate – the visitor to Gore's site could *endorse* an idea, but not *contribute* one; could agree or disagree, but not discuss. At Bush's campaign site the 'chat forums' section linked to audio clips at the 'George W. Bush Radio' page of www.broadcast.com. Similarly, the ALP's petition page is mainly about registrational interactivity, about capturing our feedback on an issue we didn't propose ourselves.

Let's imagine that the gains for a party in hosting such a forum might be outweighed by the costs – staff would need to be paid to moderate and maintain the discussion areas, and responding to messages could be time-consuming. Even this excuse, though, doesn't entirely explain the reluctance of parties to open themselves up to communication. Australian Prime Minister John Howard doesn't even have a public email address. Visitors to his site can fill in an electronic form to submit feedback, but are warned not to expect a response.[51] A similar welcome greets potential correspondents to the British Labour Party or the Scottish Nationalist Party, among others, while in Ireland, Sinn Fein sets a new standard by offering no electronic contact details at all – any communication has to be by post.[52]

The second aspect of the broadcast paradigm of web production is the application of entertainment forms – games, particularly, but also simulated TV shows. In the 1998 federal election the ALP site gave heavy emphasis to its 'Howard's End' and 'Creamfields' games, where we could kill time by pelting an image of the Prime Minister with virtual cream cakes. In the US, the Democrats offered the 'Bush Stump Speech Search Engine', a novelty item which tallied how many times George W. Bush had recycled any particular soundbite during his campaign. Meanwhile, the Republican site countered with video clips in which a simulated newsreader fronted mini-bulletins highlighting 'Gore gaffes', while Al Gore's own site urged visitors to 'watch an ad!' Back at the Democratic National Committee, visitors were offered a list of eight things they could do to help the party, including – at number six – downloading Democratic wallpaper for their computer desktop.

The nadir of this entertainment approach to date has been former Victorian Premier Jeff Kennett's campaign site in the 1999 state election. Even its address – www.jeff.com.au – traded on the politician as a *celebrity*, and on the site itself, the campaign chose not to emphasise either Kennett's record or his policies. Instead, it emphasised an online Grand Prix racing game. Also prominent among the photos of Kennett playing with his dogs was a banner promising that another game would be online soon – an electoral strategy game. But after he was turfed from office by the voters, his site quickly disappeared, denying us the pleasure of comparing his actual strategy with the promised game.

So why do these parties make so little effort to exploit the full potential of the Net – particularly its potential for creating a space for debate? One school of thought maintains that politicians *don't get it*. In this analysis, politicians are part of the generation that Negroponte termed 'the digital homeless': a generation out of the digital loop, having completed their educations decades ahead of those who now take the Net for granted.[53] One communications scholar, for example, writes that New Zealand political party sites display 'a lack of knowledge of the potential of the medium'.[54] In a similar vein, European political scientists point to the minimal use of discussion forums in Swedish and Dutch political party sites and lament 'missed opportunities'.[55] One US commentator, examining the 1996 US election sites, decides that those sites lacked 'an appreciation for the interactive nature of the nets' (sic).[56] And a UK analyst describes political party websites as 'dull and propagandistic in ways that defy half a century of learning to be more sophisticated about PR'.[57] Reinforcing these views are the occasional mocking news reports of world leaders' computer illiteracy – that Jacques Chirac was overheard asking what a computer mouse was; that former

Japanese Prime Minister Yoshiro Mori was only introduced to email in June 2000; or that Tony Blair still doesn't use it.[58]

This is a cynical perspective, which suits the mood of the times. But there's room for another suggestion which is more cynical still: these parties *get* the concept of an open Net just fine – they just *don't want it*. A major political party, after all, is a communications machine of formidable sophistication and accomplishment, equally adept at crafting ads for national TV as at setting up local telephone trees to organise door-knocking campaigns. Political parties have always adapted their strategies to exploit developments in the communications environment, from Franklin Roosevelt's radio 'fireside chats' through Bill Clinton's playing saxophone on *The Arsenio Hall Show*, and on to Gore and Bush sending themselves up on late night comedy shows.[59] Parties tend to be early adopters of information technologies, from using databases to target specific demographics, to using mobile phones and pagers to enable senior figures to manage and control campaigns at the micro-level, wherever they happen to be.[60] It seems naive at best to imagine that their broadcast-model websites are the result of ignorance. More likely is that these sites represent deliberate attempts to preserve the advantages of the broadcast model, which parties have spent decades learning to exploit to its fullest potential.

The major parties are working to develop Version 2.0 websites – sites which are interactive in the jukebox sense, but which allow no space for intercreativity. They use Jensen's transmissional interaction pattern in their adaptation of models from broadcast media, and specifically entertainment media (the simulated ads, the games, the videos). They use the consultational interaction pattern insofar as, like all websites, they enable us to interact with documents in our own time (and we shouldn't knock this facility). And they use the registrational pattern in their recruiting and capture strategies – sign up for our mailing list, make a donation, join our party – and, like many other sites, in their use of cookies to track usage. But at the level of the conversational dimension, of facilitating discussion, debate, person-to-person interaction, they're just not there.

Jeff Kennett's campaign site points to some of the drawbacks for politicians who take such a rear-view mirror approach to the Internet. On the surface, there are obvious advantages to applying expertise with television to the new media environment. But it may also have heavy costs. Political parties may be adept at using old media, but the ways in which they do this contribute to much of the voter apathy and disengagement endemic in Western democracies. Communications scholar Joshua Meyrowitz, for example, shows how politicians erode their own authority by submitting to the personalisation of television: the closer we come to our leaders, the more disappointed they make us. Television, for Meyrowitz, blurs the line between a politician's onstage and backstage personae. And while Meyrowitz attributes this to the properties of the medium, it's worth noting that it's also a feature of how politicians choose to exploit that medium. In designing its website around Kennett's backstage persona, in trading on his image, personality and celebrity status, the Kennett campaign lost its bet that this is what voters want in the Internet era.[61]

What voters *do* want in the Net era is a tough call. But it's clear that many of us don't want what we've been getting. The volatility of some voters from one election to the next is complemented by the apathy and disengagement of others: 12 million people re-elected Tony Blair in 2001, for instance, but 18 million didn't bother to show up. How much of this is due to the media's treatment of politics? In this specific instance, we might note that the UK electoral commission attributed this turnout, the lowest since 1918, to disengagement with *the campaign itself*, rather than with politics more broadly.[62]

Sociologist Manuel Castells points out that, for most people, politics takes place largely within the space of the media: 'outside the media sphere,' he writes, 'there is only political marginality'.[63] Politicians, then, must present themselves in ways acceptable to the media: for TV, images take precedence over words, and people themselves are the simplest images. But an image needs to be effective as well as simple, which draws people to the use of *negative* personal images – the politics of scandal, smear and gaffe; the politics which, Castells contends, contributes to the spread of voter disengagement.

Click here to emote

This disengagement offers a veneer of attractiveness to the idea of 'direct democracy' – the click-here-to-vote world whose proponents pitch it to us by promising that we can bypass politicians. It also contributes to the rise of political outsiders, from Ross Perot and Jean-Marie Le Pen to Pauline Hanson. What these figures have in common is their willingness to say things that are unsayable within the collusive media–politics system in which both sides know and abide by the rules. How does the Internet enable such figures to disseminate their message? As we'll see, both advocates of direct democracy and political outsiders are keen on the liberational rhetoric of Version 1.0, but are not markedly more committed to its practice than are the established parties.

Here, for instance, is former Clinton adviser Dick Morris describing his online venture, www.vote.com. 'It's a kind of online Tiananmen Square,' Morris told me. 'It's a way in which the public can speak out directly about its ideas and about the issues of the day.' Morris adds that his project allows people to take part in politics on a daily basis, rather than just on election days. He suggests that it's a forum for people from countries where free expression is constrained. And he says it offers a counterpoint to the unelected organisations that dictate the financial markets and global policy. 'This gives the average person an opportunity to manifest their views,' argues Morris, 'and be a check and balance on these global bureaucracies. The financial markets already have a way to speak out; this will be a way for the people to speak out with a voice that might eventually become just as compelling and just as loud.'

Seductive rhetoric. So seductive that it's easy to overlook the fact that Morris is really talking about market research. We're all familiar with the bored voice on the phone, asking us to indicate on a scale of one to five our degree of agreement with a statement on which we have no opinion whatsoever. But few of us would think of an

Internet version of such surveys as 'a kind of online Tiananmen Square' – even if we granted Morris that such a thing might be desirable. Morris's contribution to Net culture is his idea of the 'Fifth Estate', which he defines as 'a sort of committee of the whole, made up of all citizens online'.[64] This Fifth Estate ups the ante for the opinion poll – rather than providing data for the media, Morris sees polls as *replacing* those media.[65] He writes that this Fifth Estate is leading the US away from Madison's model of representative government, backing into the future of Jefferson's direct democracy.

This future will be reached through online opinion polls of the type hosted at Morris's vote.com site. Participants register a simple yes or no to questions on topical issues, free from such distractions as context or content. 'Should the financial records of Federal judges be posted on the Internet?' for example. Well, who knows? What's at issue here? The site doesn't really say, offering, as Morris readily acknowledges, 'much less detail' than even the sketchy *USA Today*. And most of us would need more than that to offer an informed opinion on such vote.com topics as 'Should John Hinckley be allowed to leave his mental hospital for unsupervised day-trips?'; 'Do you believe government sharpshooters fired at Branch Davidians during the Waco raid?'; or the economically phrased 'Repeal the law requiring low-flow toilets?'

As participants click their yes or no, Morris and his company pass these votes on to others. 'When you vote,' he explains, 'we email your individual vote, your email name and your opinion to your Congressman and your Senator, and also to other significant decision makers, like the World Trade Organisation.' When a vote is over, they also pass on the total, including a district-by-district breakdown for lawmakers. But what's missing from those figures? Just about everything – nuance, context, reasoning, and the identity and social situation of the person casting the vote. The value of this kind of poll was nailed by a parody on the ABC's *Future Exchange* site: 'Vote in our online poll. Is your name Bob? Yes? No?'[66] So when vote.com offers simple yes/no questions, framed in adversarial terms, and with next to no information about the issues, are those taking part registering their *opinions* or their *emotions*?[67]

'Well, they're allowed to do both,' says Morris. 'And yes/no is how Congress votes, or the MPs vote, so I don't see why we shouldn't vote that way.' But this, of course, sidesteps the idea that we elect politicians as our representatives, and charge them with grappling with all the context and complexity of issues on our behalf. We don't ask them to vote off the top of their heads. Morris is here voicing the deep cynicism with representative democracy which underlies all arguments for direct democracy of the kind offered at vote.com. The defects of direct democracy are clear from the site, and from similar projects, such as the Australian site Public Debate.[68] Both the questions and the results are skewed heavily towards some specific demographics. The person motivated to register at vote.com is still most likely to be a tertiary-educated professional whose world is reflected in vote topics such as 'Should the government break up Microsoft?' The luckless politicians who receive vote.com's thousands of forwarded emails are getting a self-selecting statistical sample that even the least scrupulous researcher would think twice about.

Is direct democracy empowering? It's certainly empowered Dick Morris, who uses the site to harvest participants' email addresses, in order to later enlist them in separate commercial surveys. In this key sense, then, vote.com is all about registrational interactivity. And we may say good luck to Morris, but his claims for the Net are important, given the column inches his standing as a former White House insider commands. Morris argues that the Net is becoming a new media force – his so-called Fifth Estate, out-evolving the battered ideal of the press as Fourth Estate. The Net user, Morris argues, as a member of the Fifth Estate, has access to the very things that the Fourth Estate can't offer without giving up its power: 'diverse information and diverse opinion'.[69]

Yet for all his use of such tropes of Version 1.0 rhetoric, Morris is proposing a very Version 2.0 ideal, sketching a future in which governance will be conducted online by private enterprise, without government participation. 'Private Web sites like Vote.com,' he writes, 'will provide the ballot boxes. *Financed by advertising*, these nongovernmental means of expressing voter opinion, in effect, mean *the end of a government monopoly* on the process of registration and voting' [emphasis added].[70]

So even government, apparently, needs a dose of market discipline – an idea that must represent some kind of end point for the neo-liberal, economic rationalist project. But while few have proposed privatising politics in this way, Morris's cynicism about political systems is by no means unique.

One Nation

> Pauline [Hanson] knows nothing about campaigning. She's great at working the room but she doesn't know where the room is. (David Oldfield[71])

Former Hanson sidekick David Oldfield is not alone in pointing to One Nation's apparent cluelessness in campaigning and dealing with the media. Journalist Margo Kingston, who covered the whole of Hanson's 1998 federal election campaign, details many examples of this in her book *Off The Rails*. 'Hansonites were a bunch of hare-brained amateurs,' Kingston writes, 'without cash, talent or experience surrounded by a marauding media.'[72] Characterising the campaign as 'a road movie without a director', Kingston describes how Hanson's organisers would set up lecterns for speeches so that they faced away from the audience – and how Hanson would deliver the speech that way anyway. She reports that Hanson, on one occasion, did an interview with the BBC believing it was the ABC, and how the ABC had to handle the sound recording for One Nation's own election commercial, as the One Nation crew were too inept to bring the right equipment.

If Kingston is even half-right, then One Nation's judgement about the media was hopeless – but they did *have* some judgement, however much they struggled to act on it. In particular, they recognised the role of the media in creating the kinds of discontent which fed the party. And they were successful in orchestrating events which let the media portray themselves in the worst light. The spectacle of police being called to eject journalists from One Nation's campaign launch was precisely the kind of event the party thrived on – such events enabled it to construct and demonise 'elites'.[73] One Nation was less the vehicle of a coherent agenda than a lightning rod for discontent. To say that they lacked policies is not to make a value judgement – they didn't even attempt to articulate positions on many of the issues which form the central platforms of the established parties' campaigns. But this was not so much their electoral undoing as it was the clearest statement of their essentially reactive, resistance identity.[74] And this resistance was directed as much against the media as against the established parties.

It's now conventional wisdom to locate the source of the discontent tapped into by One Nation within the closed circle of politics-as-usual.[75] But it's important to note that they also exploited widespread mistrust of the media. The party may have been novices in dealing with the established media, but they knew enough to realise that there were votes to be gained in people's mistrust of those media. And they were adept at using the Internet in their campaign. One Nation's 1998 federal election campaign site, in contrast to the those of the established parties, presented itself as a narrative of struggle. Visitors were greeted with the words 'You are on the web page the media do not want you to know about'.[76] This was, in fact, a consistent theme of the enormous, sprawling site, with frequent reiterations of the claim that the site 'contradicts what the media tell you about Pauline Hanson's One Nation' or that the media 'blatantly have and still do misrepresent Pauline Hanson' (sic).

Scott Balson, at the time One Nation webmaster, was the person most responsible for the Hansonites' savvy tactical use of the Net. One example of this came shortly after the 1998 federal election, when he used his email lists to organise a very large turnout of hostile supporters at an academic conference held to discuss One Nation's character and momentum.[77] In his post-election round-up, Balson went so far as to claim that Hanson had 'made history in cyberspace' and was somehow 'the first cyber-politician on the

Internet'.[78] Looking back on his own work on the campaign, Balson wrote that 'the One Nation web page... has forged the way for a new political tool'. This is a claim which would surprise, among others, Bill Clinton's staff, who incorporated email releases into his campaigns as early as 1992, and were quick to establish a White House website.[79] Leaving aside the hyperbole, though, it's one of the many ironies of One Nation that a social movement rooted so deeply in conservatism and nostalgia should simultaneously make so much of their use of new technology.

Balson was the ideal candidate for Hanson's webmaster job. Operator of his own web design business, he also ran an online 'newspaper', *The Australian National News of the Day*. Balson's newspaper was run on a paid subscription basis (this later got him into trouble with the *Sydney Morning Herald*, when he posted their stories on his site). 'It became very much a participative site,' he says, 'with people sending me email, reading the page, passing comments.' Those comments were on similar issues to those expressed by Hanson when she won the seat of Oxley in 1996: 'So,' says Balson, 'we basically had a core of people who said, hey, this woman is talking like we are.'

By the time of the 1998 federal election, Balson had had around eighteen months to develop the site as a focal point of the organisation's overall media strategy. He began by developing a mailing list and distributing electronic newsletters. Using his list, he established an inter-branch online network, sending out press releases and updates directly to branches by email. While many of their voters may not have been online, the internal organisation of their campaign was undoubtedly aided by these email releases.

'As the election came up we sent them out daily,' says Balson, 'because we were finding that the media totally distorted the message. So we had this big mailing list and sent out the press release unabridged, unedited – no Murdoch man coming in and cutting something out that's relevant or saying "well, we're going to focus on this because we can make them look nasty".' As this comment indicates, the website was not aimed at promoting *policy* so much as *image*; an aim it shared with the major party sites.

As well as paying attention to image and spin, and unlike the major parties, One Nation also experimented with the conversational dimension of the Net. In the 1998 election campaign they were the only party to feature a forum on their campaign site, broken up by topic into discussion threads; Hanson herself took part in some real-time chat sessions with members. During the 1998 campaign, I spent two weeks lurking in this One Nation forum, following two discussion threads, on multiculturalism and Aboriginal affairs. But while there were plenty of posts, not much discussion took place. Somebody would post an opinion and somebody else would post an opposing opinion and nothing much would be done with them. There was never much sense of people actually progressing towards something, or of any actual debate.

I don't make this point to dismiss the participants – they spent a lot of time doing that to each other – but rather to point to some of the limitations of the conversational dimension in enabling debate. The Net is not a soft option – for one thing, all the difficulties of offline debate also appear online, and we shouldn't pretend otherwise. There are also some new problems, specific to the Net. Recognising differences of opinion may be one of the most important objectives of debate, but in order to build on this, some form of legitimation or consensus is necessary. And, as Manuel Castells argues, 'on-line politics could push the individualization of politics, and of society, to a point where integration, consensus, and institution building would become dangerously difficult to reach'.[80] Balson acknowledges this, but in any event sees the benefits of the forum in terms of registrational interactivity. Balson's view on the role of the Net is different from the established take on cyberdemocracy. Most analyses come at this from the perspective of empowering the citizen. But from One Nation's point of view, the real benefits were for the *organisation*.

'What it actually does,' says Balson, 'is behind the scenes – it gives the key players the opportunity to see what people are thinking. And that's really the important thing.' He points to the major parties and their use of polls. One Nation lacked the resources for extensive polling, but they did have their self-selected forum participants. 'Just imagine how valuable the opinions of people randomly on a forum are,' Balson says. 'You've got all these different opinions which represent groups of people, who all have their own opinions. And if you look at those, you start getting a very clear picture for what the general feeling is in the community. And often that's very different from what the politicians believe it is.'

In the One Nation forum entrenched attitudes were voiced but there was little evidence of synthesis or development. While views broadly pro and anti the stated topic were close to equal in number, both were dwarfed by posts which seemed to have little relevance to the topic under discussion. This is a phenomenon that is common elsewhere as well. These off-topic posts illustrated the familiar tendency for Internet discussions to let the topic slide in favour of bickering, flirting, mutual congratulation, or, in the case of one post to the 'multiculturalism' debate, a picture of Itchy and Scratchy from *The Simpsons*. Much of this comes down to the desire by members to police the boundaries of their online community; it's worth noting that a very high proportion of these One Nation messages concerned the real name of a frequent poster known as 'Anonymous', whose identity occasioned more impassioned discussion than any of the actual content did.[81]

But much of this is also inherent in the limitations of the conversational dimension as a vehicle for reasoned debate – limitations which we need to be aware of if we are to make the most of its positive potential. There is, for instance, a rhetorical tension intrinsic to the hybrid nature of such posts: on the one hand they are written texts and as such may be accorded a degree of authority; but on the other hand they have much – perhaps more – in common with the protean, unresolved nature of speech. Brian Connery argues that as readers we generally think of written texts as the end product of carefully reasoned thought rather than as impromptu contributions to a dialogue. 'When we speak in conversation,' Connery writes, 'the tacit assumption is that we can change our minds about what we've said. When we write, the very act of writing seems to imply that our minds are made up.'[82] David Resnick offers a similar view: 'Conversation is not a finished product, but an ongoing process'.[83] The conversational component of online discussions, then, may limit the ability of their participants to achieve consensus. So what good is it?

There are two kinds of answer we could give to that question: one sceptical, the other more guardedly generous. The sceptical line would emphasise all the obstacles to an online discussion generating real debate: there was no evidence in One Nation's forum that anyone's views were being changed or even, at times, actually heard. This again may be an inevitable consequence of the medium: James Knapp[84] has argued that online posts emphasise the personal. Rather than following the traditional rhetoric of political discussion, posts derive their authority (or not) from the claimed personal experience of the poster. In this analysis, I'm swayed by your having lived it, rather than by your rhetorical finesse or command of details (which parallels, incidentally, a now conventional explanation of Pauline Hanson's electoral appeal). Here, again, Castells's warning about the individualisation of politics is important.[85]

A related argument in the sceptical line about the limitations of discussion forums is that while the 'freedom' from the constraints of the traditional spaces and rhetorical conventions of written argument and public debate may seem liberating, there is an additional difficulty: the Internet's nature as a public medium is compromised by its own organisational structure. Newsgroups, for instance, traded wider public access for a narrow hierarchical address and a narrow topic focus which members will work hard to police. While a web-based forum such as One Nation's could be seen as more public than a newsgroup, it's still true that corralling participants into threads of discussions held in subscriber forums means such discussions are 'public' in only a very limited sense.[86] So what wider influence can such a forum hope to have?

The answer to that question is the more generous conclusion. For the record, I'm horrified by One Nation. But however bad I think their ideas are, I also think their website in the 1998 election was very much a part of the Version 1.0 Net experience. For me, it's an important example, because it shows that Version 1.0 won't be everyone's utopia – instead, it's an ongoing struggle and series of challenges. An open, Version 1.0 media space is one in which everyone can participate, and that means having to deal with everyone's ideas, even when they turn out to be ugly, divisive and cruel. So one thing that the One Nation media experience shows is that an open media environment might turn out to be a confronting one for many of us. But that is still, I think, preferable to a closed, Version 2.0 system built around choices which all support the same entrenched positions and interests.

McKenzie Wark has argued that the microscopic nature of online discussion, complete with its bickering and flaming and its highly specific focus on issues others might find trivial or arcane, is potentially the most influential force for change that the Net has to offer. Wark writes that such discussions mark the slow emergence of a new kind of public life, in which citizens choose what its concerns and realities will be by defining them for themselves. Online politics, suggests Wark, 'means the creation of new publics, the reinvention of the polis'.[87] Certainly, the One Nation discussion forum was, in Wark's word, microscopic – although the site claimed more than 500 forum members, visitors were required to register before gaining even read-only access, and the actual number of posters was significantly lower. And yet, multiplied by the myriad other groups and forums, it may well be true, as Wark argues, that this means many people are discovering politics who otherwise might not have, finding in the Net opportunities to recognise themselves and recognise others.

Coming up

If the Internet is going to enable a new public sphere, a new model and level of discourse and participation, then on the evidence to date it's not going to be through the Net use of established political parties. Nor, if the Net is to make possible a space for politics to be revitalised, will this be achieved by duplicating the politics of the established media. Instead, the real action is taking place elsewhere. In the next chapter we'll look at a number of campaigns that illustrate *intercreativity* – using the conversational, horizontal potential of the Net to create alternative media.

CHAPTER 3

alt.media

24 March 1999: NATO Secretary-General Javier Solana gives the order for the bombing campaign against the Milosevic regime to begin. Within four hours, the Yugoslav Telecommunications Ministry suspends the transmissions of Belgrade independent radio station B92, confiscating its equipment. Editor-in-Chief Veran Matic is taken into custody. He is detained for eight hours, and forbidden contact with his family or lawyers. Matic is later released without being charged, accused, or even questioned. It is the third time in its ten-year history that B92 has been banned. But the station switches to the Net and its broadcasts are bounced back over Eastern Europe by satellite, bringing its message to much wider attention than if they'd been left alone. The Serbian authorities, says Veran Matic, simply didn't understand the possibilities of the Internet.

'Every authoritarian regime,' Matic says, 'regardless of the technical opportunities, shows this kind of attitude toward technological innovations. We acquired an Internet licence in the period when state services practically were not aware of the Internet's existence as such. Basically, they underestimated their own people... [By the time] we developed our Internet program to the extent that we obtained 50,000 daily visits on our RealAudio stream, as well as an entire network of links and mirrors, it was too late for the regime to do anything serious.'

In both their original status as an independent radio station and their subsequent Internet use, we can think of B92 as an example of *alternative* media. The sense in which I'm using the term 'alternative' here derives from the work of communications scholar John Downing. For Downing, the term denotes 'politically dissident media that offer radical alternatives to mainstream debate'.[1] In his writings on alternative media, several key characteristics of those media emerge: first, they are independently owned and managed; second, they articulate viewpoints which are in some sense dissonant from those of the wider media; and third, they foster horizontal linkages between their audiences, in contrast to the top-down, vertical flows of established print and broadcast media.

These features of alternative media are best thought of as tendencies rather than as ways of classifying – not neat labelling devices, but potential approaches available to any media outlet. Seen in this way, the concept 'alternative' becomes more useful. For example, it counters the co-opting of the term 'alternative' by commercial brokers of 'cool'.[2] Much of the music press, for instance, while keen on labelling themselves 'alternative', are wholly dependent on advertising and the recycling of record company press releases as news, which means they have much in common with commercial media, despite their publications often being distributed free.

The flip side of this is that established media organisations can sometimes show 'alternative' tendencies. In 1999, when genetically modified (GM) food was a huge issue in the UK, *The Guardian*'s coverage of the issue had a lot in common with the characteristics of alternative media: the voicing of dissonant political perspectives; the encouragement of horizontal linkages between readers; and calls for direct action. In fact, *The Guardian*'s online coverage of the story had a distinct activist slant: alongside the archived reports, a section headed 'What Can I Do About GM Food?' offered external links under subheadings such as 'Join a Protest Group', and 'Buy Organic'. These linked to groups ranging from Friends of the Earth to the Cultural Terrorist Agency, as well as to a comprehensive list of organic suppliers both online and offline.[3]

While the concept of alternative media is usually framed around questions of independence and ownership, these are issues which are reframed by the Net – the barriers to Net access are still significant for activists in many societies, but they're in a different league from the hurdles involved in setting up, say, an independent TV station. So this chapter focuses on Downing's other key points: dissonant views and horizontal connections. We'll begin with some examples of how groups are using the Net to break open the media consensus on what is and isn't news. We'll then go on to look at three key campaigns – B92, Free Pacifica and McSpotlight – which make effective use of the potential of the Net to enable new connections and coalitions, and new horizontal communication flows. Each of these campaigns is Version 1.0 in action, and each has led to the creation of a new media space. Another thing to note is that in each of these cases, a crucial use of the Internet is to attract attention from other media. As one Belgrade activist told *Wired* during the struggle against Milosevic, 'Without CNN, no doubt we would fail. That the media is here is one of our best successes. The Internet has helped get them here, and bring us more attention.'[4]

Dissonant perspectives

As an alternative mediascape, the Net teems with dissonant voices. They can articulate positions too extreme or even repellent for the mainstream media to touch. It's not a rewarding trip, so if you've never been to the home page of the Westboro Baptist Church of Topeka, Kansas, for instance, you can draw confident inferences about its content from its address – www.godhatesfags.com. The important point is that such voices can challenge the mainstream media's consensus about who and what should have access to the public sphere. The classic example is web journalist Matt Drudge and his site The Drudge Report. Drudge may not have done the world a huge favour in breaking the Bill and Monica story after *Newsweek* spiked it, but the subsequent media bonanza is a powerful example of how independent media can leverage themselves into cracks in that consensus.[5]

Careful Net use offers at least the potential to unmake the media consensus on who and what should have access to the public sphere.

Often, for instance, we see stories that are themselves about that media consensus. In Australia, a 1999 TV advocacy ad in favour of voluntary euthanasia generated one such debate. In this case, while there appeared to be a degree of agreement that pro-euthanasia views could be heard, such views could hardly be described as representative of much media discourse. So the debate in this instance centred around what was an appropriate timeslot for the ad – should it be shown late at night? Maybe mid-morning? When was the best time to screen the ad in order to minimise the chances of anyone actually seeing it?

Meanwhile, on the web, the DeathNET site is available 24 hours a day, 7 days a week, selling suicide how-to literature and pamphlets advising on the respective merits of tricyclic antidepressants and plastic bags.[6] Despite its colourful name, DeathNET is not sensationalist and has won a number of awards as a health site. The point is that while its contents lie outside the lines drawn by the established media, its organisers' Net use creates a space for these discordant voices to be heard; the site is an instance of alternative media.

Another example involves Australian government attempts to overturn the granting of a certificate to Adrian Lyne's film *Lolita*. In this case, debate turned on what the film's supposed 'effects' might be, and whether representing an act implies condoning it. In the media debate, no one came forward to present an actual paedophile's point of view – and the prospect remains an unlikely one, to say the least. It is, in fact, difficult to imagine a perspective more dissonant from that of the media consensus, so strong is the taboo – and fair enough. Yet on the web, organisations such as the North American Man/Boy Love Association (NAMBLA) are openly lobbying: 'NAMBLA's goal,' reads their mission statement, 'is to end the extreme oppression of men and boys in mutually consensual relationships.'[7] In offering this example I'm not condoning NAMBLA's views. Instead, what I want to note is that NAMBLA's project is not circulating on some Undernet of anonymous, remailed lists, but is out there on the web, taking full advantage of its public potential and its alternative media capacity – its challenge to the consensus of the established media.

Are alternative media just marginal trivia? The answer is no, but to see this we need a historical perspective. In his 1995 essay 'Alternative Media and the Boston Tea Party', Downing traces the roots of dissident publications back to the revolutionary pamphleteers of the American War of Independence. And he sets contemporary examples in a long historical context, from abolitionist writers such as Frederick Douglass up to the Deep Dish TV Satellite Network; from the nineteenth-century women's press and the suffragettes to the civil rights movements of the 1960s. He discusses the labour press of the late nineteenth and early twentieth centuries and the radical newsreels and documentaries of the Depression era.

What's important about such historical surveys is that they relate alternative media to the key struggles of their times, struggles which we can now look back on as being defining and shaping events. What Downing's work shows is that those media had some kind of impact. Each of those struggles also manifested a need for communication and information which was not being met elsewhere. Many of us might dismiss independent media as insignificant because of their size, but a better index of their potential is the degree to which governments take them seriously – a point which was illustrated by the Serbian government's closure of Radio B92.

In the realm of the censors

By the time of the 1999 crisis, B92 had already established a strong reputation within Europe. Originally a music station with a fondness for Teenage Fanclub and Sonic Youth, it was to become, by 1991, an integral part of the first major protests against the ethnification of social and economic problems in the Balkan region. As part of the broader independent media scene in Eastern Europe, B92 wasn't so much unique as a symbol for many other independent media groups, located as it was in the heart of Europe's most savage conflict in 50 years.

As B92's staff were to discover, Slobodan Milosevic understood the potential of alternative media voices very well. His media savvy extended not just to the capturing of state-run media as propaganda outlets, but also to the consistent banning and harassment of independent media. Even before coming to power he'd been involved in sacking editors of a student magazine, and under his regime independent newspapers were closed and their editors jailed. Once the war in Bosnia was underway, Milosevic removed thousands of unionised workers from the state-run Radio Television Serbia.

In March 1991, a demonstration in support of demands for media freedom turned into a riot and B92 was shut down for the first time. Allowed to resume broadcasting on condition that it played only music and issued no reporting, the station responded with a heavily coded music policy, sending out a barrage of the most oppositional records its staff could find, from the Clash's 'White Riot' to Public Enemy's 'Fight the Power'.[8] In 1992, they even banned themselves – in a tactic to draw listeners' attention to both the strength and the vulnerability of the station, B92 pre-empted a feared ban by broadcasting for a day as though it had been taken over and turned into a state mouthpiece.[9] When streaming audio became possible, the station was an early adopter, webcasting from 1996. Late that year there was an escalation of student movements and demonstrations, which led to the authorities taking out B92's transmitter. So B92 switched to RealAudio, in what was to become a powerful symbolic statement of independence. The station was really banned again in December 1996 – but this time they were ready to switch their broadcasts to the Net, on a greater scale than before.

'We started as an Internet provider when this aspect of communication was totally new,' says Veran Matic, 'except within the narrow, expert circles. That made our approach at the same time educational, propagandistic and progressive. Right from the start, we had an idea to make the Internet as accessible as possible to the majority of people, but we also wanted to use it as a serious support to our programs. (I must emphasise that it was not only news programs, but also programs from the cultural field, from civil society, publishing, etc.) We started with RealAudio stream at the moment when our transmitter was disrupted for the first time – in December 1996. From then on, we began to develop this type of transmission more decisively, which represented the base for founding the network of independent local radio stations all over Serbia. During 1997, we established the network which, due to the use of Internet and satellite, ... [became] an invincible fortress of independent news distribution.'

B92 was assisted in setting up its Net service by Dutch service provider XS4ALL ('access for all'). XS4ALL was The Netherlands' first ISP, and in 1994 co-founded the Digital City (Amsterdam) project, which for years offered an excellent example of the Net Version 1.0 in its commitment to open access and attempts to foster civic discourse online.[10] One of the key organisers here was media activist and Internet critic Geert Lovink. A former editor of *Mediamatic*, Lovink is also a co-founder of the Next Five Minutes conference series and the nettime media theory mailing list. He would also be a key figure in the Help B92 Net campaign.[11]

'B92 became famous almost overnight,' Lovink says, 'because they started to make use of this technology in such a powerful way, having the possibility to route around the authorities. Then they decided to build up a whole network, so in 1997 and 1998 they connected 40 stations and the signal went from Belgrade via Amsterdam to London and from there on a satellite, and it was bounced back all over Serbia. So all these 40 stations were able to rebroadcast these newscasts. To send out the original newscasts in Serbian and English via RealAudio is a powerful thing. When the bombing actually suddenly started [in 1999] the first response of the Serbian authorities was to pull the plug, to take off the radio signal. So B92 immediately switched to the Net.'[12]

When the 1999 military campaign threatened the survival of B92, an enormous global effort was mounted in support of its 80 full-time and part-time workers, and of the independence of the station's message. Geert Lovink points out that in such cases 'what counts is not the absolute number of listeners, but the fact that people can maintain an independent position and can bring out the message, and it can then be spread via rumours or via people printing out'.

The Help B92 campaign used the Net to co-ordinate financial, technical and moral support from around the globe.[13] During the UN embargo, Dutch authorities gave XS4ALL permission to enable a Net link between Belgrade and Amsterdam, streaming B92 live online. The signal was also picked up and rebroadcast to Serbia by the Voice of America. Real Networks, makers of the RealPlayer audiovisual software, made technical infrastructure available so that the campaign could cope with the huge volume of traffic attracted by the publicity – with high-profile links on the home pages of the BBC and CNN, the support site was drawing 50,000 hits a day, while the B92 site itself was registering upwards of 400,000. In an interview on 29 March, Veran Matic claimed that the B92 site had been visited more than 10 million times in ten days.[14] Other organisations also weighed in with support: Anonymizer.com, an anonymous remailing service, created the Kosovo Privacy Project, enabling citizens to mask their identities while sending and receiving news from abroad. Meanwhile, as the signal was being Netcast around the world, B92 journalists were shouting news reports out of the station window to whoever was down below.[15]

On 3 May, World Press Freedom Day, the international solidarity campaign called for community broadcasters to twin their station with one in the conflict zone: the emphasis here was less on financial support than on offering a point of contact and an affirmation of solidarity. A series of 'Net aid' webcasts was held – in rear-view mirror terms, these were virtual benefit gigs – with DJs and musicians from around the world, including B92 favourites Sonic Youth, taking part in '24-hour Internet parties ... to draw the attention of international music lovers to the Free B92 campaign'.[16] Donations poured in, and a 'Help B92' banner was added to hundreds of websites. Veran Matic stresses that the Help B92 campaign was important to those in Belgrade for several reasons.

'It encouraged and strengthened our position within the country,' he says. 'It showed how big was the inner support at the moment when the democratic opposition was still weak and divided into several groups. This campaign was a strong impulse for various donors and friends throughout the world that showed we were still alive and strong enough to continue fighting. The campaign meant a lot from a psychological point of view. It represented fantastic support to the B92 staff, and all those gathered around its idea. It was a clear signal that they were not abandoned in the midst of war actions, when the regime took over their station, even when the focus of various progressive political groups was shifted from the support of the democratisation of Serbia towards the refugee situation in Kosovo.'

But despite this enormous campaign, the situation for the B92 staff in Belgrade was becoming untenable, and on 2 April the webcasts were halted by the government. The station's staff were to go back on air under the new name of B2–92 in August 1999, changing their name to distinguish them from the now state-run frequency. But it would not be until the final overthrow of Milosevic in October 2000 that B92's premises and equipment were liberated and returned to Veran Matic and his crew.[17] A happy ending in some senses, perhaps, though one reached at tremendous cost. But as Lovink argues, the ability of the Serb authorities to close down even B92's Net reports throws the cyberhyped potential of the Internet into stark relief.

'We tried to maintain the independent reporting,' Lovink says. 'But that became very difficult within a few days or weeks. It turned out that even B92 couldn't maintain its independence – it was so dangerous for people to go into the building, to spread dissident voices. For the people themselves it became life-threatening, so we shouldn't really overestimate the power of the Net in that respect. If the guns are speaking, the media have to be silent.'

Far from seeing this as any kind of failure, Veran Matic points to the successes of the campaign – the station was saved, its staff stayed together, and B92 contributed to the subsequent democratic changes in Serbia.

'We could have continued for a long time with the webcasts,' Matic says, 'but it was more important for us to get back on the air, as a normal radio station. The opinion that the campaign "failed" was spread in the circles which didn't want to accept the Internet reality in the war situation, and which expected B92 to carry out some kind of technological miracle, under NATO bombs and under Milosevic's repression at the same time. In the war conditions, under severe repression and practical dictatorship, you cannot walk around with a digital camera connected to the laptop which is in turn connected to the Internet, and broadcast what's going on the city streets where the bombs are falling, or where ethnic cleansing is being carried out. You cannot do that without jeopardising the lives and safety of your journalists and activists. This is utopia. Today, it is much easier to accept that fact than it was two years ago, when the level of "mystifying" the Internet was much higher. Maybe the Internet experience during the war in Yugoslavia made us better understand its good qualities, as well as its shortcomings.'

The novelty of the Internet means its use in political actions attracts media attention. But of course such uses are not in themselves new. The consistent logic of oppositional movements in marginalised situations is to turn the decentralising potential of communication technologies against state efforts to control and centralise information.[18] We might think, for instance, of the Chinese students in June 1989 who pieced together a public address system on the Beijing University campus.[19] Or the cassettes of 1977 speeches made in Paris by Ayatollah Khomeini, which were circulating on the streets of Tehran hours later.[20] Or the 30,000 copies of a newsletter, produced by a Panamanian exile, which spread around the country by fax and photocopy as Noriega grappled to retain control of the country.[21] Or the Jordanian who used a fax machine to circulate 1000 copies of a censored article from *The Economist* around the country.[22]

Internet activists who counter government censorship are the latest line in a long message – whether it's Indonesian students using chat groups to exchange advice about resisting troops during the ouster of Soeharto,[23] or South Korean student radicals using the Net to rebuild their movement after their leaders were forced into hiding.[24] The Chinese online newsletter *VIP Reference* is another example. It bypasses the Chinese government's controls on the Net by being sent to subscribers from a different email address every day.[25]

B92, like all these groups, sought a Version 1.0 model of open communication, while the Milosevic regime pursued a brutal Version 2.0 closed system – although with a twist. In 1995 B92 had established Belgrade's first Internet service provider, receiving unexpected permission from a Telecommunications Ministry official who thought the state wasn't doing enough to establish Net services. The same Ministry which was later to ban the station had provided it with the infrastructure to get around that ban.[26]

One year on from the overthrow of Milosevic, Matic told me that B92 is in some senses no longer alternative media. Instead, it has grown into a kind of de facto national public service broadcaster. 'At this moment,' he says, 'we are trying to achieve something that is really hard: to establish balance between commercial effects of our programs and promotion of progressive and educational ideas. Those who tried to achieve this goal often failed. We are not going to let that happen.'

Free Pacifica

Government censorship, whether it involves shutting down radio stations or shutting out movies, is not, of course, the only kind. Equally insidious is what political scientist John Keane labels 'market censorship'.[27] Keane argues that the rhetoric of the 'free press' has been co-opted by corporate forces and free marketeers. These forces invoke the Fourth Estate ideal of liberty of the press in the service of deregulation, but their own practices render this ideal an antiquated joke. For Keane, to direct calls for press freedom against the spectre of *state* censorship is to miss the real dangers inherent in a *commercialised* mediascape, as media corporations become more integrated and oligopolistic. Keane points to the contradiction between freedom of communication and freedom of the market:

> Those who control the market sphere of producing and distributing information determine, prior to publication, what products (such as books, magazines, newspapers, television programmes, computer software) will be mass produced and, thus, which opinions officially gain entry into the 'marketplace of opinions'.[28]

In this analysis, the argument that market forces 'give the people what they want' is only true so long as what the people want is commercially viable. As with political party websites, active citizenship is recast as active consumerism. Keane paraphrases the logic of market censorship as 'We offer you all kinds of choices so long as you, the consumer, restrict your choices to the terms agreeable to us, the entrepreneur[s]'.[29] Market censorship – the restriction of media content to that which fits the corporate consensus – can also be seen in the commercial pressures on independent media. In 1999, B92 wasn't the only radio station to have need of an Internet support campaign.

31 July 1999: More than 10,000 people march through the town of Berkeley in California to demand community control of local FM radio station KPFA. The largest demonstration seen in the town since the height of the Vietnam War, this is the climax of a series of protests against the management practices and perceived commercial ambitions of KPFA's parent body, the Pacifica Foundation. More than 100 people are arrested during July, as armed guards protect Pacifica's headquarters. The crisis has been brewing for years, but comes to a head with a series of firings in mid-1999: general manager Nicole Sawaya is removed on 31 March, followed nine days later by award-winning journalist Larry Bensky. On 13 July, broadcaster Dennis Bernstein is sacked after an on-air scuffle as he plays material alleging that Pacifica aspires to sell KPFA, estimating that its frequency and brand could be worth as much as US$70 million.[30]

This combination of alleged profiteering ambitions with heavy-handed corporate management is anathema to KPFA's history and image. Founded in 1949, it claims to be the world's oldest listener-funded FM radio station. In terms of horizontal communication, Pacifica's founding charter enshrined the belief in communication among diverse groups, rather than the voicing of a single dissonant perspective.[31] In its 50 years on air, it has stood out against McCarthyism and played a major role in the Berkeley Free Speech Movement. It has broadcast a number of celebrated firsts, from the first reading of Ginsberg's 'Howl' and Patty Hearst's denunciation of her family as 'capitalist pigs',[32] to what, in 1958, may have been the first substantial US documentary on gay rights.[33] Writers and artists, including movie critic Pauline Kael and novelist Alice Walker, found an early outlet for their work at the station. By the 1970s its reputation was such that its newly opened sister station in Houston was bombed by the Ku Klux Klan. 'When the theater is burning,' said one Pacifica brochure, 'our microphones are available to shout "fire".'[34]

KPFA is largely funded by listener donations, and rejects corporate sponsorship and advertising. Combined with its strong community emphasis on the California Bay Area, where it claims around 200,000 listeners, this demonstrates the potential effectiveness of alternative media organisations in fulfilling needs ignored by commercial media. Besides its commitment to issues and viewpoints left untouched by the mainstream media, the station has a strong focus on horizontal communication, seen in its history of attempting to democratise decision-making and access provisions among its largely volunteer staff.[35]

But recent management strategies appear geared towards commercial principles, towards maximising audience numbers at the expense of diversity. Matthew Lasar, author of *Pacifica Radio* and an active opponent of the changes which led to the crisis, argues that the cause of these changes was decades of government policy that had undermined community broadcasting – most recently, the Telecommunications Act of 1996, which enabled corporations to own a larger number of radio stations in a given region than previously; the resulting inflated prices made local programming increasingly economically unviable, as well as inflating the dollar value of spectrum space such as that of KPFA.[36] This led to management policies that contrasted sharply with the views of the listeners, subscribers, and much of the staff: major decisions about this community radio station were increasingly taken without consulting the community itself (its listeners, its volunteers, its employees). Media critic Edward Herman argues that Pacifica's new emphasis on its commercial worth was misguided, that the notional dollar value of the station represents only 'the advertising potential of a signal when all other values are ignored'.[37] The key thing here is that large numbers of the listening community perceive the market-oriented changes – the cuts to niche programming, for instance – as damaging to those other values; this perception itself in turn becomes damaging. The crisis can be seen as a question of market censorship, of the culture of the bottom line steamrollering a home-grown culture of local journalism and community participation.

As with B92, the KPFA support campaign made effective use of the Net.[38] 'As an organising tool, the Net played a crucial role,' says Matthew Lasar. 'Dissidents could now very rapidly coordinate demonstrations, pickets and public events on a national level. And much valid information did get passed along.' And, again like B92, the basic infrastructure for this campaign was already in place before the 1999 crisis. In 1995, former KPFA employee Lyn Gerry helped set up a listserv for Take Back KPFA, a small movement of disgruntled staff and listeners. The online activity enabled new coalitions among disenfranchised staff. The list was a response to Pacifica's so-called 'gag rule' – an injunction against discussing internal matters on air. This meant KPFA staff were unable to discuss the pressure to become more commercial or what they perceived as management's union-busting tactics. But savvy Net use enabled them to bypass the gag rule roadblock.

'Our listserv enabled the first major breakthrough in garnering press coverage of the issues in mid-1996,' says Lyn Gerry. 'I had been fired from my job in late 1995 for vocal opposition to the new regime at Pacifica, and for defying orders to not speak to the press. I wrote a nine-page report and distributed [it] on the listserv. I did not even put my phone number on the report. It was not a press release. There were [only] about 100 people on the listserv. Within a week, my phone began ringing – reporters from all over the country were calling me about my report. For the first time, some reporters realised there was a story. I learned then that dropping a piece of information onto a listserv is like dropping a rock in a still lake – the ripples will expand to touch shores beyond your horizon.'

'The widespread use of the Internet in the mid-1990s,' says Matthew Lasar, 'enabled Pacifica dissidents to share information within the organisation with unprecedented speed. The results of Board meetings, incidents at stations, firings, governance decisions and news about Pacifica-related public events or gatherings all became nearly instant knowledge to thousands of people within the organisation and beyond. Equally important, discussion lists enabled dissidents to establish a common sense of identity, in part because participants in these lists often could not actually see each other, but only exchanged words. People who often had little in common now had one thing in common – some grievance with Pacifica radio and an institutional venue for expressing it, complete with an audience. As Pacifica weakened the power of Local Advisory Boards, eliminated station folios, enforced the gag rule and purged programmers, dialogue about the future of the organisation streamed into the Internet. In truth, there was nowhere else to go.'

In June 1996 this campaign added a website, providing ready-to-download posters and flyers, and aiming to make documents available to journalists.[39] As well as its dossiers on Pacifica, the site has hosted a number of online petitions, and offers contact details for supporters to get in touch with Pacifica, politicians and media outlets, as well as with each other. Besides this horizontal emphasis, the importance of these information flows was that, firstly, they weren't subject to Pacifica's gag rule and, secondly, as Gerry notes, any attempts to close down the site could have been interpreted as confirming the charges made on the basis of the 'incriminating' internal documents the site posted. By the time of the crisis in 1999, Free Pacifica had grown to become the largest site on the issue, attracting over 1000 visitors a day, with traffic eventually growing so much that it overwhelmed its visitor logging program.

Again recalling Downing's point about the importance of horizontal linkages in opposition to vertical top-down media flows, Gerry points to an unexpected outcome of the Net campaign for KPFA – the technology now made the formation of an alternative radio network possible.

'One of the interesting by-products of this,' she explains, 'was that other community radio stations became involved, either because they aired Pacifica programs or because they were facing the same battles at their stations – pressure to mainstream, professionalise and corporatise their message and operations. Because high quality audio files can now be moved across the Net, stations in far flung areas can cooperate to create an audio wire service, the type of thing that in the past only monied news organisations with satellite uplinks could touch.'

This wire service now operates as an online exchange centre where independent radio journalists and producers can upload their programs and freely download others for broadcast.[40] It's a model of Version 1.0 Net use. The activists surrounding the KPFA issues take full advantage of the Net's potential for independence, both financial and editorial. They also exploit its capacity to establish horizontal linkages and connections. Matthew Lasar hopes that these developments can be extended beyond the specific local issues of the Pacifica campaign.

'I believe that if a settlement is achieved at Pacifica,' Lasar says, 'the Pacifica campaign should redirect its direct action energies towards the real source of the problem: the corporate state and its executors in the US – the Federal Communications Commission, the Corporation for Public Broadcasting, and the National Association of Broadcasters. For twenty years these forces have been systematically neutralising local non-commercial radio of all kinds. If Pacifica is reformed, I hope that the powerful energies the Pacifica campaign tapped into to take back the network can be redirected towards the forces that have been boxing in the network for two decades.

Technology now can to some extent expand the non-commercial domain. But it can only go so far when the corporate state is so obviously committed to making marginal all forms of indigenous, community based media.'

McSpotlight

16 February 1996: Shortly after 10 am a small crowd gathers outside a McDonald's outlet in London's Leicester Square to watch a part-time bar worker and an ex-postman switch on a computer. Using a borrowed laptop and mobile phone, Helen Steel and Dave Morris, the defendants in the McLibel trial, launch a new element of the anti-McDonald's campaign: the McSpotlight website, an enormous repository of information, criticism and debate about McDonald's.[41]*A subsequent press conference at London's first Internet café generates a lot of media attention. McSpotlight's server registers more than 35,000 visitors within the first 24 hours, with severe congestion keeping many others out by the first afternoon. The administrators are to claim more than 12 million hits for the first fifteen months of operation, although a more realistic perspective on this comes from their calculation that this represents more than 300,000 actual visitors; 1700 visits in the first week alone originate from McDonald's own computer system.*[42] *Dave Morris will later pronounce McSpotlight 'another nail in the coffin for McDonald's censorship strategy'.*[43]

After defending themselves against the corporation's libel allegations in England's longest-ever trial, Steel and Morris are world authorities on most of McDonald's strategies, particularly their communications strategy. This encompasses not only an enormous annual advertising budget, and extensive community promotions, but also a lengthy trail of litigation.[44] Those who have issued apologies on receipt of sharp letters from the chain include the BBC, who apologised in 1984 over a report linking beef production to deforestation, and *The Guardian*, who kowtowed three years later, after publishing an article about the corporation's employment practices. Besides these large, well-resourced organisations, other targets have included a Scottish youth theatre group which wrote a satirical play about the fast-food industry without mentioning McDonald's by name, and the owner of a small sandwich bar called McMunchies.[45]

Steel and Morris, along with three other members of a small autonomous group called London Greenpeace, were issued with libel writs in September 1990.[46] The group had distributed a six-page leaflet titled 'What's Wrong With McDonald's? Everything They Don't Want You To Know'. It made claims about a range of corporate practices, from exploitative advertising through nutritional information to labour policies. The leaflet drew connections between multinational food chains and hunger in developing countries; it questioned the extent of McDonald's use of recycled packaging; and it argued that McDonald's ads 'deliberately exploit children'. It also attacked the corporation's treatment of animals and of its workers, and was decorated with doctored logos reading 'McCancer' and 'McMurder'.[47]

'The leaflet brought all these arguments that different people were making against McDonald's and put [them] all in one place,' explains Jessy, one of the founders of McSpotlight, who prefers not to reveal her surname.[48] 'The environmental movement, animal rights activists, trade unions, nutritionists, advertising standards campaigners: all these people were amassing evidence against McDonald's. All the leaflet did was tie them all together and ask: "do you still want to support this company?"'

Faced with the Byzantine complexities of the UK's libel laws, and potentially ruinous costs, three of the group apologised and backed out; Steel and Morris, however, angered by what they saw as corporate censorship, chose to fight the suit. What followed was a surreal expedition through the legal system and corporate power, taking 314 days of court time, encompassing a counter-suit in which the defendants turned the tables on McDonald's for distributing 300,000 leaflets of their own (which labelled Steel and Morris liars), and spawning a film and a book, as well as the McSpotlight site.[49]

'In the context of the campaign by grassroots activists against censorship and the UK libel laws,' Dave Morris says, 'the imaginative and determined efforts of those involved with McSpotlight were an additional dimension, a real boost, and an absolute winner. It was also the only way we could see of making the full details of what came out in the trial widely accessible at the time, and ever since.'

'McDonald's spend a fortune each year on advertising,' Jessy points out. 'And everybody knows their point of view. But the people campaigning on the other side had, up to this point, been limited in how far they could get their arguments heard. This was partly financial, partly practical and partly geographical. The Internet solved all those problems in one. One of the first things we did was to upload material that McDonald's had managed to suppress during previous legal actions – newspaper articles, leaflets and films which McDonald's thought they'd seen the last of... which was a great public service, I think. As the McSpotlight slogan says, you could then "judge for yourself".'

The site itself exists as an unlikely consequence of the UK libel laws. The McLibel defendants were required not only to prove that every word of the offending leaflet was true, but to do so using primary research sources – previously published materials were inadmissible, so Steel and Morris had to conduct interviews and depose witnesses. This meant that as the case went on, their research developed into a formidable archive of information. The initial idea for the McSpotlight site was to create a kind of library.

'When the trial had been going for about maybe a year,' says Jessy, 'there was this huge amount of information that had been amassed: we're talking serious research which was being used as evidence in the trial. And basically, the only way people could access that information was by going round [to] Dave's house, going in the spare room, and trying to go through all these piles and piles of papers which were just bundled in the corner.'

With more than 21,000 files taking up over 120 Mb of space, McSpotlight is potentially daunting to new visitors. But its creative design and skilful organisation demonstrate that a successful activist site must cater for participants with wildly divergent levels of interest and expertise, both in the technology and in the issues. A guided tour of the site's key features with audio commentary by Steel and Morris is available, for example. The material ranges from a complete archive of transcripts of the trial through to satirical animations; from a history of McDonald's through to more than 600 links to other anti-corporate campaigns. The McSpotlight organisers are proud of this mix of range and depth.

'If you're writing an article,' explains Jessy, 'and you just want a soundbite, one example of why McDonald's is bad, McSpotlight hands you that very easily. Or you can just pop in for some light relief with the quiz or the pages on what McDonald's executives said in court. But on the other hand, if you want to find out about bacterial cross-contamination of beef burgers as they're refrigerated in South Africa, or some ridiculously in-depth thing, that's also easily accessible. Which is why the Internet and hypertext are so brilliant – there isn't any other medium which lets you do that.'

This organisation of content into different levels is one of the strengths of McSpotlight. Information is obviously a key to successful campaigning, so one of the core functions of any activist website has to be not only making hard-to-find information available, but also presenting it in as accessible a fashion as possible. The very success of McSpotlight shows, paradoxically, that the Net is *not* a soft option for activists. Making the most of its potential demands serious skills – not in state-of-the-art design or animation, but in information management and provision. If, for example, web designers don't try to anticipate the needs and abilities of those most likely to use their site, information overload is a real risk.

As the Net continues to expand, filtering out signal from noise becomes, if not harder, then certainly more and more time-consuming. Anyone who has tried to do research online will be familiar with the hyperlink spiral, where each site opens up possibilities which hadn't been evident before: the right piece of information is always one more click away. This can be exhilarating and productive, of course, but it can also be paralysing, producing nothing more than the feeling of being in a dense swamp of bookmarks and printouts.

Whole books have been written about such info glut (thereby adding to it).[50] The scale of the problem can be gauged from a search for 'information overload' on the Google engine in July 2001 – it retrieved more than 298,000 matches. Altavista turned up 250,303,903, as well as an invitation to 'shop the web for "information overload"'. Sociologist Tim Jordan argues that there are two kinds of data glut common to cyberspace: 'First, there can be simply too much information to absorb. Second, information can be so poorly organised that finding any particular piece of information becomes impossible.'[51]

Jordan describes how cyberspace not only offers more information, but also encourages people to produce more of it: email, newsgroup posts, simple web pages. He argues that the catch-22 of this is that attempts to manage information overload invariably involve technological fixes which themselves create more information: the point-and-click graphical user interface of the web browser, for example, makes online information more accessible, but also makes it easier for still more people to generate yet more content. Jordan notes that using a software program or intelligent agent to manage or filter your email, calls or news means having to master the additional information you need to use that program effectively. And he argues that applications which offer synopses of information (stock tickers, daily news digests) contribute to information overload rather than managing it, because each synopsis interposes one more piece of potentially unmissable data.[52]

McSpotlight certainly contains more info than most of us could absorb, with everything from the full text of Justice Bell's 800-page McLibel judgement to an account of a campaign of hoax bomb threats by anti-McDonald's activists in Poland. But Jordan's second point, that information can be so badly organised as to be impossible to find, is one the McSpotlight team is well aware of.

'McSpotlight was set up as a library,' Jessy explains, 'but obviously it's not just a library, because a library is like "this is the book, this is the catalogue, it's A, B, C, D . . . ". The website makers have to make paths through it so different people can use it in different ways and get the stuff that they want, and I think McSpotlight has been fairly successful in that: there's different levels for different kinds of people who're interested in different things.'

The test here, of course, is how easily other campaigners can actually make use of the information at the site. 'For me,' says Dave Morris, 'the most important issue is to change the world, replacing an oppressive, unfair and destructive capitalist/governmental system with an anarchist society based on freedom, sharing and cooperation – this can only be done by people in every corner of the globe, especially working class people, gradually building up strong and vibrant collectives, networks, workplace groups and communities, and challenging the power of those institutions which currently dominate our lives. Will the Internet help, or will it fragment and hinder this process?'

One example of a group McSpotlight helped is Mountains Against McDonald's (MAM), in the Blue Mountains near Sydney. While MAM may not have seen themselves as working towards an anarchist society, as Morris puts it, they did offer an example of a local community networking internationally, and using the Internet to help this process.

September 1996: McDonald's applies for permission to open an outlet near Katoomba in the Blue Mountains, emphasising job creation and a projected $900,000 annual payroll. The local community mobilises to influence the local council. Their campaign centres around traditional strategies, including leafleting, letter-writing, and a rally on 8 December that draws 700 people to hear messages of support from prominent figures, including David Suzuki. Early in the campaign, MAM media spokesperson Gillian Appleton finds the McSpotlight site, and draws upon its case studies of residents' actions around the world for the two issues of a broadsheet newspaper that the group is to produce.

'McSpotlight proved to be the most marvellous resource,' Appleton says. 'We were able to put together a dossier for local councillors showing that they were not alone, that other councils had knocked back this huge corporation, and to make them feel good about doing their bit. And it put me in contact with other useful groups. I put up a message about what we were trying to do, and from then on I received many emails from sympathisers all over the place.'

On 7 February 1997, McDonald's notified MAM that it would not be proceeding with its application. In any event, the local council were to reject their application two weeks later, having accepted the campaigners' arguments about the need to preserve the distinctive identity of the Blue Mountains area.[53] In the same month, on its first anniversary, McSpotlight doubled in size, as all the official court transcripts of the trial were added to its resources.[54]

The Blue Mountains campaign is an example of activists using a medium routinely conceptualised in terms of its *global* nature, to facilitate action at a *local* level. Recognising the importance of this, other groups are turning to Geographic Information Systems (GIS) to organise data into forms which are not only accessible but can also be acted on in specific locations. These databases offer a map interface which users can manipulate to zoom in on a particular region, town or suburb to filter the relevant information. GIS is already becoming familiar in commercial applications, such as street directories for large cities.[55] If you're having a party, for instance, you can use a mapping database of your city to highlight your street and its surrounding area on a grid map, mark your address, along with the locations of the nearest train station and bottle shop, and either print out or email the whole thing along with the invitation. This kind of combination of enormous databases and simple maps offers powerful possibilities to activists.

One interesting example is the Hybrid Media Lounge, a project co-ordinated by the Society for Old and New Media in Amsterdam. This site maps individuals and groups across Europe who are active over a broad spectrum of cultural politics and digital media.[56] By using the mapping database, we can compare groups from country to country, trace their networks of collaboration and dialogue, and cross-match demographic data on funding, infrastructure and resources. The database also shows how the participants rate themselves on a range of qualities and areas of specialisation: groups from most countries, for instance, rate themselves as stronger in 'practice' than in 'theory'. Those from the former Yugoslavia score themselves highest in 'political correctness'. And almost everyone gives themselves top marks for 'intellectualism'.

In the UK, Friends of the Earth (FOE) launched its Factory Watch GIS in February 1999.[57] By clicking on their own region or town, or by entering their postcode, anyone can access data on toxic emissions from factories in their area, sourced from the government's Environment Agency. The data can be viewed in a range of different forms, allowing people to quickly build their own dossiers on particular suburbs, factories, corporations, chemicals or health risks. In terms of grassroots campaigners actually making use of this Factory Watch data, Mike Childs, a senior pollution campaigner with FOE, acknowledges that there are some problems with the database as yet, noting that the worst-offending factories tend to be located in low-income areas, where Internet access can be assumed to be low. But he points to the project's potential, and argues that the availability of the information in this form is already having some impact.

'It can already be regarded as a success in terms of influencing business,' Childs says. 'Many businesses have told us that they were inundated with calls following the press coverage of the findings. It has also influenced the UK government and the Environment Agency in terms of demanding reductions of pollutants [and] also in terms of developing improved pollution inventories. It also has the potential to increasingly influence shareholders, especially with increased research into health impacts and the potential for new developments in biology to be able to let people know which chemicals they are particularly susceptible to, [thus] leading to [possible] legal actions.'

Environmental campaigners have been building databases on toxic emissions for years: in the 1980s the US Public Data Access project collated a database, searchable by zip code, which could be used, for example, to calculate potential statistical correlations between cancer incidences and the locations of nuclear facilities.[58]

But combining these kinds of huge databases with a simple map interface offers powerful new possibilities for structuring the data into an accessible form. A clickable map was used, for instance, in the NetDay 96 project, the aim of which was to use volunteers to wire every school in California to the Internet. By zooming in and out on particular districts or schools, potential volunteers could gauge the extent of existing participation, and decide whether or not their own contribution might help.[59]

19 June 1997: The McLibel verdict is in. Chief Justice Bell rules that McDonald's exploit children through their advertising; are cruel to the animals used in their products; and pay low wages, 'thereby helping to depress wages for workers in the catering trade in Britain'.[60] He also points to discrepancies between the nutritional claims made in McDonald's promotions and the actual content of their food. Steel and Morris are, however, found to have libelled McDonald's on a number of other counts, and are ordered to pay £60,000 in damages. But, perhaps sated on bad publicity, McDonald's announce that they will not be pursuing this. A McSpotlight volunteer is in court and phones the details through – they appear on the site before the mainstream media have been able to turn the story around. Many reporters in fact contact McSpotlight to find out the verdict. Interest in the case is at its peak, and following the verdict the site is accessed more than 2.2 million times; by 1999, McSpotlight will claim to have been accessed more than 65 million times.[61] More than 3 million of the original leaflets, available as ready-to-print downloads at the site, in languages from Japanese to Swedish, have been distributed in the UK alone since the beginning of the legal proceedings in 1990, including 400,000 in the days immediately following the June 1997 verdict. The corporation's attempts to halt the leafleting, including their application for an injunction against the defendants, are abandoned.

Remarkably, given the corporation's track record of legal action and the site's inclusion of material which had previously been suppressed, McSpotlight has attracted no official attention from McDonald's:

no cease-and-desist letters, no writs. This may suggest that the corporation has learned the hard way – the case, on which they spent a reputed £10 million, and which attracted some of the worst coverage imaginable, was a PR debacle. But it could also be tacit recognition of the unfeasibility of shutting the site down.

McSpotlight is solid evidence for many of the claims that have been made for the resilience of Internet materials in the face of censorship attempts. Leaving aside the fact that a huge number of people have already visited the site and read the material, including the leaflet at the centre of the libel trial, McSpotlight is hosted on several different servers around the world. Any legal action would need to be co-ordinated across a number of different legal systems and jurisdictions – the libel provisions of at least some of these may be less weighted in favour of the litigant than are those of the UK. The entire site contents are also available on CD-ROM, and a complete earlier version of the site was for a time available as a compressed download, enabling anyone to retrieve and host a copy on their own website.

Any attempts to censor the site would almost certainly result in a PR disaster to rival the court case itself, as supporters would set up additional mirrors and publicise both their locations and the legal action. While the famous claim that 'the Net treats censorship as damage and routes around it' may be frustrating in its technological determinism, in its emphasis on the properties of the *technology*, the geographical dispersion of McSpotlight is evidence that many *people* are both able and prepared to 'route around' censorship attempts.[62]

It's significant that McSpotlight continues to draw visitors, even with the trial long over. One reason for this may be 'Beyond McDonald's', its expanding section about campaigns against other transnational corporations. But its debating room, a cluster of ongoing discussion forums, also continues to generate substantial traffic on the trial and the issues raised, more than four years after the verdict. The importance of this opportunity to discuss the issues raised by the site can't be overstated. While the McLibel defendants benefited greatly from the Net, using it, for instance, to collect witness statements by email (a significant advantage over overseas phone calls, given their lack of resources), they both remain committed to the power of leafleting. Dave Morris emphasised to the *McLibel* documentary makers that 'it's a very immediate form of dealing with basic ideas – people can stop and have a chat, they can see that you're a human being. Whereas, for example, people getting things from television or from books, there's no real communication.'[63] Taking this comment on board, McSpotlight claims to have been among the first to exploit the potential of the conversational dimension on the web, to adapt the two-way potential of early, computer-mediated communication applications to a web-based campaign.

'One of the ways that McSpotlight was pretty pioneering,' says Jessy, 'was that very, very early on we got an email from the guy who'd developed this discussion room software... [it] is now so widespread that it doesn't sound exciting, but at the time it was "let me have it!". He offered it to us for free, and so the McSpotlight debating forum came in early on. This [is the] sort of communication which McDonald's and corporations like them definitely don't want, in particular [among] the workers. Obviously they don't allow unions at McDonald's, so there's one part of the debating room which is specifically for workers, and as soon as that started, it was immediately bombarded with people saying things like "in my McDonald's we have shifts ending at 9 am", and "my wages are this" and then somebody in, say, Canada would respond, and so on. So this was definitely not just two-way communication, but communication that led to action – when they swapped stories in the debating room, then they could take that to their store and say "wait a minute...!"'

But corporations live or die by their communication strategies as much as do social change campaigners. Any innovative use of a new medium, such as McSpotlight's early adoption of discussion forum software, can easily be co-opted for corporate promotions. While McDonald's have yet to try this particular approach, Shell Oil promotes discussion forums at its website, on topics including 'Human Rights: None of Our Business or the Heart of Our

Business?' and 'Islands of Wealth – Wages and Local Investment'.[64] Unsurprisingly, a large number of posts displayed in the Shell forums are extremely positive about the corporation, and while it may well be true that there are large numbers of people out there who are keen to express their pleasure in Shell products and practices, others have shown that this is exactly the kind of support which money can buy. For example, unable to count on spontaneous groundswells of public enthusiasm, many corporations have turned to the practice of 'astroturfing' – contrived 'grassroots' support campaigns which combine corporate funding with state-of-the-art PR. The Center for Media and Democracy works to expose such astroturf campaigns, issuing a range of publications analysing groups such as the now-defunct National Smokers Alliance.[65] This pro-tobacco lobby group, with its slick website, claimed to have more than 3 million members supporting its struggle against 'anti-smoking zealots [who] have shown that they will stop at nothing in their goal to create a smoke-free society'. The catch with this campaign was that it openly stated that it was founded with cash from the Philip Morris corporation, makers of Marlboro and Benson & Hedges, adding – with no obvious irony – that its policy of 'no-strings-attached independence . . . has allowed the NSA to earn contributions from two other cigarette manufacturers'.[66]

Faced with the realities of this kind of corporate counter-activism, the challenge for campaigners is to keep ahead of corporate responses, to continually experiment with new approaches. This doesn't mean entering a software arms race or relying on state-of-the-art equipment as much as it means finding new ways of exploiting the basics – such as email – to enable genuine communication, the exchange of relevant information, and making this information as easy to act on as possible.

'I always think that small groups of individuals can stay ahead of the lumbering corporations and the big media organisations and governments,' says Jessy. 'Anything that's a huge corporate body can never move as fast as individuals. So use your brain and your imagination to come up with new ways of doing things faster, doing things funnier. Don't use what McSpotlight did – it was great at the time because it was pioneering, but it's got to be something better and more interesting and more exciting now.'

Coming up

The McSpotlight, Free Pacifica and B92 examples illustrate the potential of alternative media to open up a space for otherwise unheard viewpoints. In each of those campaigns, one result has been the creation of a new media space – the new role of B92 as de facto public broadcaster; the radio program distribution network set up out of the Free Pacifica campaign; McSpotlight as anti-corporate library and Internet hub. Each is intercreative, unfinished, Version 1.0. Each also illustrates the potential of communication between participants, rather than to audiences – something which is also central to our next major example. I interviewed Jessy in early November 1999, when she had just discovered an Australian Net project called Active Sydney.

'The next exciting stage,' Jessy said, 'is what the people at Active Sydney are doing. They've created the database and the means, just the infrastructure, but they're not creating the content. Anybody who goes to the site can add news stories or events or whatever. So the content is made by the visitors, as opposed to being a one-way thing. You don't want the web makers to be controlling the content, because then it's just the same old story, same as TV or whatever, where a few people are making the content and everybody else is just listening. So I think that's the way forward.'

A few weeks later events were to prove her right, with the establishment of the first Indymedia site, on which Active Sydney was a crucial influence. The next chapter looks at the resulting global network of Independent Media Centres (IMCs), an intercreative movement which is also trying to create alternative media and an alternative mediascape. The IMCs place the emphasis on the *production*, rather than the *consumption*, of media texts. And they stress the conversational dimension of the Net as the creation of DIY media, rather than just as a means of debating the writings of others.

CHAPTER 4

Open Publishing, Open Technologies

12 September 2000: The Opening Ceremony of the Olympic Games is just three days away and an estimated 22,000 journalists have descended on Sydney for the largest programmed media event on the calendar. During the Games, journalists will be able to draw on the resources of the International Broadcast Centre at Homebush Bay, with its 3200 staff, 70,000 square metres of TV broadcast facilities, 700 cameras and 400 video machines. And that's just for broadcasters, with hundreds of additional staff and thousands of volunteers deployed to assist print journalists.[1] But not everyone has such resources. In a converted warehouse in the inner-western suburb of St Peters, the Sydney Independent Media Centre (IMC) is also gearing up for the Games.[2] Sharing the building with an anarchist bookshop, an office of Friends of the Earth and some residents, a dozen computers line the room. Some are hand-painted or stickered, having been donated or salvaged from dumpsters. Volunteer contributors cluster around terminals, typing up stories and helping each other upload video and audio files. Spare monitors and disk drives in various states of repair are piled on shelves, jostling for space with Linux penguin dolls, an anti-uranium mining flag, and banners reading 'do-it-yourself media' and 'cause for comment'. With its scraps of taped-down carpet and mismatched chairs, the IMC resembles a garage band rehearsal space, a model of DIY culture, makeshift and make-do. The IMC, observes volunteer Gabrielle Kuiper, 'runs on the smell of a burnt-out modem'.

The Sydney IMC is a Net-based vehicle open to all to exchange information, contribute stories and discuss ideas. Initially set up to provide coverage of Olympic-related issues other than the medal tallies, it was still going strong many months after the closing fireworks of the Games. Its main feature is a web page that automatically publishes submissions from participants, who can contribute text, photos and graphics, video clips and audio files. Its database automatically updates the site to make each new submission the lead item. While the website is the virtual centre, and anyone can contribute items from home, work, libraries or Net cafés, the physical office space offers volunteer support for the more technically demanding submissions, such as uploading audio and video clips.

The IMC software represents a confluence of interests, influences and experiences which makes it, in many ways, the state of the art in Internet activism. Building on our earlier analysis of alternative media, in this chapter we'll look at two key aspects of the IMC movement: its advocacy of open publishing and its links with the open source software movement.

Open publishing is the key idea behind the IMC. There are no staff reporters as such – instead, the content is generated by anyone who decides to take part. There is no gatekeeping and no editorial selection process – participants are free to upload whatever they choose, from articles and reports to announcements and appeals for equipment or advice.[3]

'We wanted people to be active participants, not passive readers,' explains Gabrielle Kuiper, recalling the ideas behind the Active Sydney project that led to the IMC. 'They are the ones with the knowledge of their events, groups and news, not us – why should we be gatekeepers? If you come from this philosophy – basically respecting the intelligence and creativity of your fellow human beings – then good communication practice and making the site easy to use follows from that.'

Contributing a story, as IMC programmer Matthew Arnison says, is no harder than using Hotmail – type your article into an online form and click on a button to submit. And in an important use of the conversational dimension of the Net, each story includes an option that allows others to add their own follow-up comments, so that each story can be the catalyst for an online discussion as well as a standalone item. Before the IMC was even officially launched (on 7 September 2000), more than 300 items had already been posted; over 120 people had subscribed to the main mailing list; and the site was receiving around 1000 hits per day, with most of this traffic coming from word-of-mouth, incoming links and the existing global network of more than 30 other IMCs.

The IMC network is a key example of the Internet Version 1.0 in its commitment to openness – not just to open publishing, but also to the open source software movement. While the predecessors of the IMC software – discussed below – could be described as backing into the future, in that they essentially updated Usenet newsgroups for a web environment, the addition of video and audio facilities means that an IMC now extends the advantages of the text-based, conversational dimension into the multimedia environment of the web. The developers are also aware of the potential pitfalls of backing into the future, arguing, for example, against live video streaming in favour of clips – streaming seems the obvious application precisely because it's most like what we're most used to: television. Well-chosen clips, in contrast, enable us to access the relevant information whenever we want, rather than having to log on at the precise time of a webcast.[4]

The first IMC was established in Seattle for the World Trade Organisation events of November 1999. In the ten months following Seattle, a network of more than 30 such IMCs had been set up, each using the same freely circulated software, and each relying on individual participants or visitors to submit content. The Sydney site was only one of five new IMCs to go online in September 2000. By March 2002 there were more than 70. IMCs have been established for one-off events, such as May Day in London, and as part of longer-term, localised political campaigns, from India to the Czech Republic, from Italy to the Congo. The Brazilian IMC, for instance, offers ground-level analysis of trade issues in a choice of three languages, while the Israeli site offers eyewitness accounts of conditions in the West Bank and Gaza.

This dispersed network is a key to thinking about the opposition to globalisation which the contributors to IMCs tend to share. First, we should note that the 'anti-globalisation' label is reductive – globalisation is a complex matrix of processes, but the content of the IMC sites tends to address opposition to global *corporations* rather than to *all* aspects of globalisation. Second, it's entirely consistent that this opposition to dispersed global capitalism is itself dispersed and localised, rather than centralised. If corporate power is everywhere and nowhere, nomadic and dispersed, then opposition to that power needs to be likewise.

The routine emphasis on the global nature of the Net means it's easy to overlook the significance of *local* applications of computerised communications. In fact, activist groups have been capitalising on this potential since well before the Internet took off in the mid-1990s. In the early 1980s, for instance, community groups in New York established computerised databases to calculate the risk of arson attacks by landlords intending to claim on insurance policies, number-crunching such variables as a landlord's fire history, tax arrears, and record of building code violations.[5]

By the late 1980s, local government and civic network infrastructures were establishing computerised presences, and homeless people in Santa Monica used local library connections to the city's Public Electronic Network to mount a successful campaign for improved access to showers and lockers, both essential in finding work or accommodation.[6] In Wilmington, North Carolina, the residents of the Jervay Place housing project used public access Net terminals to secure a more active, participatory role in its proposed redevelopment. Through discussion lists, they made contact online with architects who analysed the redevelopment plans (sent to them as email attachments) and offered advice, which the residents then took to negotiations with the housing authority.[7]

'While globalisation has had a massive impact on the distribution of ideas, culture and technology,' says Gabrielle Kuiper, 'people are still geographically based. The vast majority of people cannot fly to London for a protest, although the WTO protests in Seattle in November 1999 were a dramatic example of mobilisation across the US and Canada. Not only will there always be transport restrictions, but many of the structures that affect people's quality of life are geographically based. And despite the rise of transnational decision-making institutions such as the European Union and the World Bank, my view is that most vital decisions are still undertaken by the governments of nation states.'

This local dimension to Net activism can often be misrepresented as a model in which innovations flow from a perceived 'centre'. The IMC movement is an example of this, with Seattle sometimes seen as the ground-zero of the project. For instance, in a feature about anti-Olympic activism, *The Guardian* in the UK reported in July 2000 that Sydney sites were 'modelled on those used to plan and publicise the protests in Seattle'.[8] In fact, while the Seattle IMC was the first to be established, the software source code which these sites use originates in Sydney.[9]

The IMC format is a development from a local site called Active Sydney.[10] Like the IMCs, Active Sydney is an open publishing forum, though it has an additional emphasis on co-ordinating direct actions and discussing tactics. Gabrielle Kuiper, a PhD student at the Institute for Sustainable Futures, first had the idea for this site at two o'clock one morning while she was supposed to be thinking about her research (although she stresses that creating Active Sydney became very much a collaborative process and evolved through the input and ideas of the site's volunteers). The initial inspiration was a monthly photocopied listing of local social change events, the 'Manic Activist' calendar. Compiled by student environmentalists, the calendar listed direct actions, lectures, seminars, meetings and screenings.

'As a compulsive reader of noticeboards, all-round information junkie and email addict,' Kuiper says, 'it was obvious to me that this sort of information should be online. While it originally focused on events, Active is attempting to be more than a subversive newspaper or a community bulletin board. It aims to be a meeting place, an online autonomous zone, a hub of active information where a whole variety of social change movements connect. What we hope is that participants can connect their talk to action, either by discussing events on the calendar or using the discussion to facilitate future events or actions – even things as simple as using the forum to write a joint letter to a politician or corporation or bank.'

Kuiper had an extensive network of connections to activist groups in Sydney, including to Critical Mass,[11] but the technical expertise needed to bring them together online was co-ordinated by Matthew Arnison, also a PhD student, researching physics at Sydney University. Arnison was a co-founder of the Community Activist Technology group (CAT) in 1995 – CAT's orientation can be seen in its slogan 'pedestrians, public transports and pushbikes on the information super hypeway'. Among other projects, CAT offers training and geek help to those who want to take progressive causes online;[12] in helping to devise Active Sydney, Arnison was able to bring this experience to that of the two programmers who'd begun the project.

In his landmark essay on open source software, 'The Cathedral and the Bazaar', Eric S. Raymond suggests that 'Every good work of software starts by scratching a developer's personal itch'.[13] The itch that was to lead to the IMC software was a frustration with the unproductive hierarchies and group politics of other alternative media. Like many in the city, Matthew Arnison had been unhappy when the previously Sydney-focused Triple J radio station went national, leaving a community gap which smaller local stations, such as 2SER, struggled to fill. Missing the underground feel of the old Triple J, Arnison gravitated towards community TV, but became frustrated by management and top-heavy group politics and moved on to the Net. With CAT he built connections with independent groups working in both old and new media, developing software to enable online coverage of an Australian community radio conference in 1995; the group's first piece of automated software was developed for the 1997 conference. This software enabled continuous updates from correspondents at the event, extending information and participation to those who couldn't attend. But Arnison again became frustrated by hierarchical management, an experience which would be pivotal in the development of the horizontal Active software.[14]

Active Sydney first went online in January 1999, using most of the features of the subsequent IMC software – open publishing, email alerts and events calendars.[15] By March 2000, the site hosted contact details for more than 100 organisations in the city – community TV and radio groups, organic food co-ops, public transport activists, disability rights discussion lists, sexuality collectives, legal advisory services, local branches of Amnesty and Friends of the Earth, and Ecopella – 'a community group that performs modern choral works about environmental issues'.[16] None of these groups receives much attention from the mainstream media. But what's perhaps more significant than this exclusion is that on the rare occasions when they do gain coverage in those media, that coverage will be framed along very predictable lines. Actions and events will be shaped to fit the familiar *protest* genre. Conflicts and oppositions will be highlighted or manufactured, and discussion of issues will be replaced by a depiction of disruption to the status quo. The issues and causes that generate any political action are inevitably marginalised to fit the narrative pattern of 'protesters clash with police'. Here's how the next stage of the IMC story might look if written in this genre:

18 June 1999 (J18): As the G8 leaders assemble in Germany, a wave of protests, co-ordinated in cyberspace, breaks around the world. Dozens are arrested outside the New York Stock Exchange, while hundreds of activists form a human chain around the Treasury Department in Washington. London sees its biggest riots in years, as police turn water cannons on a crowd of thousands that has torched a bank and trashed a McDonald's. In Nigeria, an estimated 10,000 demonstrate against oil companies, and others take to the streets from Uruguay to Nepal.

This kind of protest narrative emphasises events at the expense of processes, effects at the expense of causes. And it ignores the more interesting story of the J18 event – the global DIY networking and the formation of new coalitions and alliances between disparate and globally dispersed groups (and globally dispersed programmers). The J18 actions were the precursor to Seattle N30, and to the subsequent wave of co-ordinated events in Washington, London, Melbourne, Prague and Genoa.

On the day of J18, Sydney was in the first time zone to go live, and the Active team had revised their software to include video and audio facilities. This mix of text, sound and image, fundamental to the web, recasts a whole raft of long-standing problems for activists. Communicating to a potentially large audience has previously meant grappling with the logistical difficulties of access to newspapers, radio and TV. In principle, the Active software could now function as all of these and more.[17]

'For decades,' says Matthew Arnison, 'activists have had their work cut out working against centralised media. Not only in terms of the rise of media monopolies, but also in the technology itself. Even in community radio and TV stations, the centralisation of control that having a single transmission tower implies does create big problems for diverse and activist media. With the Internet going mainstream, we have an electronic media where the corporations are on the back foot – not just because the technology is new, but because the technology inherently supports decentralisation, many-to-many and two-way communication. Instead of activists having to subvert a centralised media technology, it's the corporations madly trying to subvert a decentralised technology, and so far largely failing.'

As plans for the Seattle actions took shape, Arnison met some American activists who were planning a site to cover the events. They envisioned an online newswire for alternative media outlets, such as community radio stations in the US, but were planning to use commercial software to build it. Arnison persuaded them that the technical advantages of a site based on free software would make international collaboration more possible. The Active Sydney software offered the basis.

Working from Sydney, Arnison was instrumental in getting the Seattle IMC online from scratch in the two weeks before the events.

As the protests began, physical distance from Seattle turned out to actually be an advantage: Arnison was able to focus on programming and systems administration problems more clearly than could the geeks on the ground, who were stretched by helping contributors digitise their video footage and post their stories. It also meant he didn't have to deal with the tear gas which intermittently wafted into the Seattle IMC office.

DIY media and news narratives

Tear gas wasn't a problem at the Sydney Olympics. But when the Sydney IMC was launched, some form of direct political action at the 2000 Games seemed a certainty. After all, every previous Olympics of the satellite broadcast era had seen dissident events of some kind, from the Black Power salutes in Mexico City through the massacre in Munich to the bomb in Atlanta. This inevitability is not just due to the media attention: just as major sporting events are designed to focus nationalist values, they also inevitably focus dissident voices. In trying to draw a nation together, it is natural that those who feel excluded will try to add their voices. Global events such as the Olympics, which trade in metaphors of tribal conflict and war – just as politics trades in sporting metaphors – cannot be separated from the political.[18]

Many political issues were flagged in advance as directly implicated within the Olympics, including the environmental impact of the Homebush development; the corporate practices of major Games sponsors in the light of the rhetoric about the 'Green Games'; and the broader social impacts upon renting tenants and public transport users. The Games were also seen as a platform that could highlight the stalled Aboriginal reconciliation process. And for a time there was even talk of large-scale protests against new legislation designed to prevent protests. While there were, in the end, no protest events at the Sydney Games on the scale that many had predicted, the IMC may prove to make a more lasting contribution to such political issues than demonstrations would have made. It's a formidable achievement in its creation of a space for people to voice their views and hear the views of others, and its fostering and expression of a do-it-yourself aesthetic.

In this sense, the IMCs are part of a long tradition. Downing's historical surveys of alternative media enable us to trace this tradition back hundreds of years. But a key root of the IMC approach is in the tradition of the fanzine (now generally abbreviated to zine). In his punk history *England's Dreaming*, Jon Savage writes that 'Fanzines are the perfect expression – cheaper, more instant than records. Maybe THE medium. A democratization too – if the most committed "new wave" is about social change then the best fanzines express this.'[19]

If the sound of punk was fossilised by the time the Sex Pistols released their second single, the more lasting impact was in the spread of the DIY aesthetic typified by the zine – the rise of indie record labels and distributors changed music more than any of the records they released; and, as Savage suggests, the boom in zines enabled by photocopiers opened up new possibilities for self-expression. Mark P, of seminal punk zine *Sniffin' Glue*, claimed he would write each issue in a straight draft with no revisions, with the final copy including crossed-out lines. 'I didn't really care about the magazine,' he told Savage, 'it was the ideas that were important.'[20] There's a direct line from early zines to the open publishing philosophy of the IMCs.

This access principle – here's three chords, now form a band – is the central pillar of what cultural studies academic George McKay terms 'DiY Culture'. McKay defines this as 'a youth-centred and -directed cluster of interests and practices around green radicalism, direct action politics, new musical sounds and experiences'.[21] The contributors anthologised in McKay's collection *DiY Culture* flesh this out, with their participant accounts of setting up indie newspapers and video magazines, of Reclaim the Streets and Earth First!, of free parties and traveller culture. Central to all of these is the circulation of ideas: this is not just a precursor to activism, it is activism – writing *is* action, open publishing *is* a direct cultural intervention.

Stephen Duncombe, in his book on zines, confesses to an anxiety that the cultural space of indie publishing is a retreat from actual engagement, from physical action: DIY media, in this analysis, are only 'a rebellious haven in a heartless world'. What good is all this subcultural creation, Duncombe asks, if it remains 'safely within the confines of the cultural world'?[22] The current state of development of Net-based activist media, as represented here by the Sydney IMC, offers an opportunity to rethink this separation of writing and action, of commentary and participation.

Questions about alternative media are usually framed in terms of their political economy, around issues of ownership and access; the DIY impulse of the online zine lets us think of alternative media in terms of *content*, in terms of the ways in which that content is framed and presented. The need for this becomes clear when we look at the ways in which established media represent extra-parliamentary political action.

In the first two weeks of August 2000, the *Sydney Morning Herald* and *The Australian* printed a total of eight stories about possible protests at the Games. The actual issues which would have occasioned any such protests were never mentioned. Instead, each of these stories was framed as a simple *narrative of conflict*. The potential protesters were depicted as agents of disruption, as a possible danger. One report said that 'Anti-Olympic activists are being urged on the Internet to block the entrance to the Games opening ceremony and disrupt it with sit-ins'.[23] *Activists, urged, block, disrupt*. On 7 August, the *Herald* disclosed that British 'anarchists' were 'likely to hit Sydney for Olympic protests', and had plans to 'ambush the wives of corporate bosses'.[24]

The underlying causes of these possible protests were not misrepresented or distorted – they were simply omitted. How can we explain this? Media academic Todd Gitlin describes the routine journalistic approach to narrative construction as: 'cover the event, not the condition; the conflict, not the consensus; the fact that "advances the story", not the one that explains it'.[25] If we understand conflict-based narratives as constructions of events, rather than as neutral accounts of them, we can understand the sameness of the news stories discussed above, and the reasons for their excision of the issues behind the possible protests. So one possible explanation is that this kind of coverage is a consequence of journalistic reliance on tight narrative formulas. The standard news story, of course, like other kinds of story, needs narrative drive; needs character, place and action; needs to link events in causal sequence. Above all, to enable characterisation and cause-and-effect, it needs *conflict*. Issues are personified to create characters who are then depicted in conflict with one another, with the attribution of quotes creating a virtual dialogue between these actors.[26]

Stuart Hall has argued that in telling a story, the tendency is to tell it the way it has been told before, drawing on 'the whole inventory of telling stories, and of narratives'.[27] From this perspective, the stories cited above from both the *Sydney Morning Herald* and *The Australian* represent formulaic recourse to a familiar narrative pattern, with the issues implicated in any extra-parliamentary political action subordinated to the simple conflict-based schema of 'police clash with protesters...'; a consequence of what one commentator has described as the 'daily creation of nonfiction drama'.[28] As communications scholars Elizabeth Bird and Robert Dardenne have suggested, the conspicuous similarity of media accounts of the same events should be understood less as evidence of the virtues of 'objective' reporting, and more as a 'triumph of formulaic narrative construction'.[29]

One challenge for Internet activists, then, is to develop ways of telling stories which are issues-focused, without replicating the conflict-based narrative structures of the established media. Hall has pointed to this trap:

> I often say to radical friends, 'I'm not interested in what the person's politics are; what kinds of stories do they tell?' Because I know many radical journalists in the media who tell exactly the same stories: they construct events with the same kinds of language as the people who disagree with them profoundly.[30]

The point here is not to criticise *all* kinds of narrative. Film scholar Robert Stam has argued convincingly that much of the pleasure of news is in fact derived from its narrative structures, that stories are pleasurable because 'they impose the consolations of form on the flux of human experience'.[31] But one opportunity presented by new media vehicles such as IMCs is rather to frame issues in ways that still entail these 'consolations of form' without recourse to reductive formulas of conflict and disruption.

The opportunity – and the challenge – for open publishing is to find new ways of writing which bring audiences closer to solutions to the problems under discussion. Stories that address complexity rather than reducing it to a good guys/bad guys schema. Stories that stimulate discussion and debate rather than constructing conflicts. Stories that go beyond a spurious objectivity and recognise their writer's responsibility to strengthen civic discourse and involve community members in coverage of issues which affect them.[32] Stories that are part of an intercreative, unfinished, Version 1.0 media space – a messy space, but one in which people collaborate in making their own DIY culture, rather than selecting from the news-jukebox of prepackaged points of view.

To a certain extent, the content of the Sydney IMC resonates with such objectives already, although it is still early days. Contributions range from reports on events to appeals for equipment; from participants' accounts of actions to complete reports and media releases from NGOs and lobby groups; from announcements of upcoming events to photo series. There are also often postings of links to stories in the mainstream media (although the posting of established media source material does raise the question of whether this simply re-legitimises those media as the authentic forum for news). With no gatekeepers or editorial staff, the content is entirely supplied by members of the community. Some write in traditional journalese, some rant, others cover every point in between. This kind of open publishing forum creates opportunities and challenges for its participants: one challenge, as I've said, is to develop new ways of reporting – a more open, Version 1.0 approach which doesn't simply replicate the adversarial style of the established media. This will take a while, if it happens at all, of course, but the opportunities are there to explore ways of writing about issues and events other than the conflict narratives that we're all so used to.

But beyond the level of the individual story, the accumulating content of the IMC can also be seen as an alternative narrative in itself. Just as each 'protesters clash with police' story contributes to a larger 'us against them' narrative in which forces of order are ranged against agents of disruption, so each IMC item becomes part of a broader narrative. This larger narrative, comprising all the articles, clips, appeals, links and comments, may be contradictory and inchoate. But for those involved, it has the potential to address and represent the complexity of their social contexts, rather than reducing them to a consoling form. As I've suggested already, the Net is *not* a soft option for activists. If it reframes the problems of ownership and access, it also opens up new challenges, such as the opportunity to move beyond conflict narratives. All this new infrastructure is cool, sure, but to expect that new infrastructure to solve our problems for us would be to succumb to a kind of technological determinism.

Getting determined

On my desk is an issue of *Newsweek*. As media coverage of technology goes, its cover story is entirely typical – 'How the Internet Is Changing the World'. 'The modern world has gone digital,' *Newsweek* tells us, 'and there's no turning back.' Inside, the contents page promises that the magazine 'takes you inside the revolution and explains how e-life is changing all of us'. [33] *How the Internet is changing the world*... behind this idea is a complex web of assumptions and ways of thinking about technology. In the popular media, we still read and hear about how this or that technology is *going to change things*, apparently all by itself.

This perspective, in which technology is somehow autonomous, an independent realm from which new products appear by surprise and then change everything, is a common one.[34] But as technology historians Merritt Roe Smith and Leo Marx point out, this type of narrative depicts technology purely in terms of artefacts and their 'effects'. 'It is noteworthy,' they write, 'that these mini-fables direct attention to the consequences rather than the genesis of inventions.'[35] The *Newsweek* issue is a typical example of the gee-whiz story in which a machine will change things whether we like it or not. Such stories help create the popular sense of technology as an independent causal agent. But they fail to acknowledge, as technology critic Langdon Winner reminds us, that 'deliberate choices about the relationship between people and new technology are made by someone, somehow, every day of the year'.[36]

We're so used to *Newsweek*'s kind of story that it seems counter-intuitive to think of technologies as having in-built political properties. We've seen the bumper sticker so many times – 'Guns don't kill people, people do'. But to assert in this way that technologies are neutral involves a degree of sleight of hand. And, as we'll see below, the Independent Media Centre movement is entirely dependent upon a wider movement of Version 1.0, technology-based activism – the open source software movement. But to make this connection, we need to take a few points separately.

First, we need to recognise that technologies are shaped by particular people in particular contexts. Science philosopher Nelly Oudshoorn demonstrates how the implementation of the contraceptive pill was highly dependent on context: so much so that there is still huge resistance to it in many societies. Oudshoorn argues that public acceptance of the pill depended on a number of social factors, including the existence of adequate medical infrastructure; a high level of acceptance of prescription drugs and medical controls on the individual body; and a sexual climate capable of accommodating the negotiation of contraception between women and men. Clearly, not all societies meet these requirements and, as Oudshoorn drily notes, 'many of the required transformations were beyond the power of the inventors of the pill'.[37]

Second, we need to note that the uses to which a technology is put are not determined by its technological properties alone – those uses are determined by the interaction of social, cultural and economic forces, and the end result will reflect competing interests and objectives. The sealed receiver of the contemporary radio, for instance, is very different from its original form as a point-to-point communication technology. The telephone was originally used as a broadcast medium – particularly in Hungary, where the Telefon Hirmondó network delivered news, drama and music programming over the phone system from 1893 until 1925, when it merged with the state radio broadcaster.[38] Meanwhile, Thomas Edison initially thought his phonograph would be used for office dictation.[39] The point is that it was not just the intrinsic properties of either radio or the telephone or the phonograph that determined what we came to use them for. Their present uses – and the modified forms in which they are now sold to us – reflect processes of societal shaping which manifest themselves in what we can call *in-built politics*.

Examples of in-built politics of technologies are all around us. The medium may not be the message, but there *are* messages built into each medium. Not just those messages that we think of as *content*, but those that are embodied within the *form* of the medium itself. Messages which reflect the values, assumptions or agendas of those shaping the technology. For example, film scholar Richard Dyer describes how photographic media and cinematic lighting have been developed to favour the filming of white people – on the assumption that this was where the money was – to the extent that within the film industry 'photographing non-white people is typically construed as a problem'.[40] Dyer traces how photographic innovation has customarily used the filming of the human face as its benchmark, and taken white faces as the standard. He describes how experiments with lighting, aperture size, development times, and the chemistry of film stock 'all proceeded on the assumption that what had to be got right was the look of the white face'.[41] Technologies of photography, then, come with in-built politics.

To understand the open source movement we need to recognise that in the worlds of computing and the Net there are also in-built technological politics. This was the serious point behind Umberto

Eco's playful claim that the differences between the Macintosh and DOS computer operating systems were analogous to a religious schism. The Mac interface, Eco suggested, was Catholic, with its ease of use making it possible for all to enter the kingdom of spreadsheet heaven; the difficulty of DOS, by contrast, was a more demanding faith, which assumed that some would not make it all the way.[42] The Mac interface certainly took many people to spreadsheet heaven who might otherwise not have made the trip. In his history of Apple computers, Steven Levy enthuses about their desktop visual metaphor, describing it several times as 'brilliant' – 'What better way to emulate the sort of work that most of us do with computers – deskwork – than by making a virtual desk, with virtual drawers, virtual file folders, virtual paper?'[43] And indeed, in terms of popularising computers, an interface that the initial corporate target market could quickly get to grips with was an undeniably smart idea. But as for the rest of us? Whether we're using our machines at home, or at school, or on the beach, we're also still using that same office metaphor — a Version 2.0 interface.

But while the Version 2.0 corporate imagery remains dominant on desktops more than 20 years later, the web adds an extra layer of metaphor. In commercial web design, a common current interface is based on the newspaper front page. For some Internet critics, such as Geert Lovink, this is 'a regressive move, back to the old mass media of print'.[44] In such an analysis, the newspaper metaphor conceives of the Internet *participant* as an information *consumer* – we may read newspapers in many different ways, but we don't generally have many opportunities to contribute or interact with them, beyond letters to the editor. But for Matthew Arnison, the combination of newspaper imagery and the open publishing potential of the web is reinventing the idea of the newspaper. Not as a medium of consumption, but of intercreativity.

'Most people spend most of their web surfing time making their own newspapers through email, egroups, instant messaging, and personal web pages,' he argues. 'I see Indymedia's identification of open publishing as part of a broad pattern of how people use the

Net. Time and again, parts of the Net that take on aspects of open publishing prove to be enormously popular. Geocities is in the top ten websites, and the big thing about Amazon is the community of book reviews it fosters – Amazon has a strand of open publishing at its heart. It's interesting to me that we are moving from a highly *corporate* model for computer use, the cubicle, to a much more *community* space – the newspaper. When you turn on most people's computers you get a corporate desktop, but within seconds the Net whacks a funky wallpaper over the top. The Net is the space where people are seriously messing with the way we interface with computers. And it's feeding back.'

In other words, the shaping of technology is not just down to governments and business, but also to the rest of us – we can *adapt* technology as well as *adopt* it; the best description of this remains novelist William Gibson's observation that 'the street finds its own uses for things'.[45] We might think, for instance, of hip-hop's adaptation of the turntable, turning an instrument of reproduction into one of new production. Or of early house music producers' use of the Roland 303, taking what was designed as a home practice aid for guitarists and turning it into the creative centre of an entirely new style of music.[46] Such examples cause problems for a number of the most common assumptions which have featured in cultural responses to new technologies – among them the hard determinist notion which Langdon Winner, in criticising it, summarises as follows: 'technological innovation is the basic cause of changes in society and . . . human beings have little choice other than to sit back and watch this ineluctable process unfold'.[47] In fact, the open source software movement – which underpins online open publishing in both technical and philosophical senses – shows that many people spend every day proving this inevitability to be false.

'This is why the Internet and the free software behind it is great,' argues Arnison. 'There is a reason for this inherent decentralisation in the technology. The tools were designed by geeks who wanted to chat with other geeks, not by Microsoft to maximise their profits, and really not by the military either, despite their role in funding it.

The geeks didn't want bottlenecks, editors, censorship. [This decentralisation is] in the very low-level workings of the Internet system. To "fix" this, global corporate giants [would] need to get control of enough of the hardware and software that runs the Internet, and change the rules.'

When Microsoft suddenly noticed the Net in early 1994, this looked, in Arnison's words, 'like a snack'. With its existing dominance of the PC software and operating system markets, it appeared all too likely that the company could use these virtual monopolies to gain another. But the emergence of the Linux operating system and the corresponding wider interest in free software – or open source – has dented this scenario.

When Arnison talks about 'free software' he doesn't just mean applications which are available without charge, although this is often the case, but applications where the actual programming code is openly accessible rather than masked, so people are both allowed and encouraged to adapt and customise it for themselves. More than this, the software itself cannot be privatised for commercial exploitation: it can be sold, but only if buyers are allowed to continue the modification process, should they want to. This stimulates innovation, experimentation, improvements, which are then passed along for others to build upon. New developments are continually being tested and either built on or rejected. Among other consequences, this means that open source software is extremely reliable, continually going through a kind of peer review process. It's free, as the saying goes, as in free speech *and* as in free beer.

'I find this whole process a total inspiration,' says Arnison. 'A lot of it hinges on the idea of copyleft, where you include a GNU Public Licence with the source code, which means no one else can ever privatise the software. It's like a virus that enforces caring and sharing on people who use your software. All of the tools we use [in IMCs] are free software, which means anyone has the right to use, copy, or change them. This includes access to the source code, the blueprint behind the software, which means we can customise it or add features with wild abandon. Having reaped the benefits of free software, pass them on.'

This idea of non-proprietorial invention is not a new one. Salk refused to patent his polio vaccine, and Benjamin Franklin left his many inventions unpatented, arguing that 'we should be glad of an opportunity to serve others by any invention of ours, and this we should do freely and generously'.[48] And it's important to underline that this is far from a fringe movement – Matthew Arnison points out that the open source Apache web server has double the market share of any competitor, including Microsoft. Apache has even been used to run Microsoft's own Hotmail.[49] And key IT corporations, including IBM, Netscape and Hewlett-Packard, have embraced the principles of open source software to greater or lesser degrees.[50] In February 2002, *New Scientist* magazine published an article about developments in this area which extend the principles beyond computing, from OpenLaw to OpenCola. But perhaps the real significance of that article was that the magazine issued it under an experimental licence, encouraging readers to copy, redistribute or modify the text – the article itself was open source.[51]

Such open source software is not only a powerful armoury for activists – it's an activist movement in itself.

'Free software,' says Arnison, 'is the main resistance against the privatisation of the Internet. This is a huge struggle for the future of mass media. I like to think the free software movement is a very strong activist movement, but somewhat hidden. This is because the people involved don't really think they are activists, and other activists don't realise what is going on. The organisation of the movement is highly decentralised, with people all over the place spontaneously starting new projects, and placing them out there to see if people like them. One big reason the people making free software don't realise they are activists is because they are having so much fun! For once they can write software to do what they want, rather than what they can get paid to write, and they get to join a huge family of other people sharing the process of writing software.'

The IMC philosophy of open publishing is, then, entirely consistent with its technical foundations in the open source movement. Both essentially argue that anyone can and should be trusted to be both creative and responsible. And both are the Version 1.0 Internet in action. In yielding editorial control in favour of relying on participants to be responsible in their contributions, the IMCs trust that a self-selection process will keep the projects on track.

'If you provide a space,' says Gabrielle Kuiper, 'where people can be intelligent, imaginative and creative, they will be. We've had no problems with people writing inappropriate or even boring news, for example. If you look at the news stories, you'll see that a lot of thought and effort has gone into most of them.'

The open source and Independent Media Centre movements are not, of course, the only Net projects with a stake in the in-built properties of technology. At the level of cultural activism, some interesting Net art works to raise awareness of the fact that our sexy new tools are not neutral. One example is Word Perhect (sic),[52] a web art project which displays a crude, hand-drawn word processing interface. We are invited to choose a surface to write on – the back of a phone bill, an old calendar, scraps from a cigarette packet – and then to select either messy or tidy handwriting. While the icons on the toolbar are mainly familiar, clicking on them produces messages which call attention to our increased dependence on certain kinds of technology to shape our writing, our correspondence, our thinking. Select the spell-check button and you're told to go and look the word up in a dictionary. Ask to open a saved document and you're told 'you should close your eyes and remember about it yourself'. And the 'undo' button urges you to take responsibility for what you've just written.

What's this all about? It teases us about our reliance on automated functions, but it also points to the in-built values in word processors, to a potential gulf between the automatic functions which toolbar programmers decide we want, and the reality of the writing process (Word Perhect includes a kettle icon for tea breaks). As Jeanette Hofmann argues, learning to write with a word processor involves learning a kind of new language.[53] Word Perhect draws our attention to this and to the degrees of skill and autonomy that programmers allow us.

A more sophisticated project than this, though, is Natural Selection. Natural Selection is a search engine mind-bomb, created in response to racist, white supremacist material online.[54] It works to undermine the supposed objectivity of mainstream search engines, interrogating the ways in which they select and rank their results. As Matthew Fuller, one of the creators of Natural Selection, argues, 'The apparent neutrality of these technologies and their in-built cultural biases need to be taken apart'.

Created by members of the Mongrel art collective in the UK, Natural Selection offers a conventional search interface. But if we enter certain search strings, we land in a parallel network of content created by Mongrel and their collaborators. The idea is that anyone searching for, say, neo-Nazi writings would find themselves instead at a site which ridiculed their views. But the engine is not only activated by searches for what could be considered offensive key words. Let's take 'cats' as a harmless example. When we run our search, the engine retrieves a list of matches to cat-related sites – all of which are aliases for links to Mongrel's own content. Click on our cat choice and we might find ourselves reading a self-assessment questionnaire to gauge our chances of being accepted for immigration into the UK. Fill it in, and an article appears about Jamaican woman Joy Gardner, who died on 28 July 1993 after being forcibly restrained with a 'body belt' while she was being issued with a deportation order. Other destinations include an article about neo-Nazi rock bands by post-Situationist author Stewart Home and an essay about Islam and globalisation by Hakim Bey.

The result? Nothing returned by the search engine is what it seems, nothing can be taken at face value, and we may be left to question the processes by which our usual search mechanisms select and organise their material. Matthew Fuller argues that the project points to the in-built cultural politics and assumptions that regular search engines exhibit; that their methods of selecting and presenting results mask intrinsic biases towards the world views of the kinds of users that their programmers imagine. There may be no way around the use of such assumptions, of course, but one point of Natural Selection is not to criticise programmers for doing this, but to ensure that we are aware that they do it when we go to use a search engine.

'The authority of the mainstream search engines is derived from the relative accuracy of their ability to crawl and order the Internet,' Fuller says. 'There are a variety of ways in which they do this. Yahoo!, of course, trades on having a legible view of the world, a way of ordering things and dividing them up into neat little definitions. Let's just say there's not much poetry going on here. This is the Internet made into a filing cabinet. For the other two techniques, when their crawler hits a search engine, they organise what they find in different ways. Either a fixed view of where words or contents fit, or a system of ordering that is determined according to the choices users make about where they go from a list of search results. The first is really a hidden, automated form of directory; the second is more subtle, but seems largely to be determined by associations of search terms with search results determined by whatever is the demographic majority of Net users. This has obvious implications.'

Natural Selection is also significant because it engages with one of the central scare stories attached to the Net – the claim that information found online just can't be trusted. It's a common argument – that there's no central quality control, no peer review, and no editorial selection. This is a typical journalistic response to the open publishing model of the IMCs. In fact, it's striking how often those articulating this position are those with a stake in older media and in those media's self-appointed role as the Fourth Estate: journalists who, trained to work in certain ways, for certain kinds of institution, want to convince us (and maybe themselves) that those ways and institutions work better than they in fact do. This despite the fact that it's only too easy to find fault with the established media, from the specious veneer of objectivity to the blatant cross-promotions. Anyone who consumes enough media can write their own list.

In his history of the Internet, John Naughton – himself a journalist – reminds us of the established media's less-than-spotless record on addressing Net issues, reviewing *Time* magazine's legendary 'cyberporn' beat-up, in which a crude undergraduate research project was brandished as evidence of the need for censorship in cyberspace.[55] Perhaps, then, to appropriate some corporate-speak, the status of information on the Net is not so much a problem as an opportunity. Perhaps the IMC open publishing model offers – or demands – a new way of looking at older media: a Version 1.0 lens through which we can view Version 2.0 media. Rather than just reminding ourselves not to be too trusting of what we see online, perhaps we should extend this scepticism into a more active engagement with other media forms as well.

Coming up

In their commitment to open publishing and open technologies, the IMCs are a model of the Internet Version 1.0. And they're a model of contemporary alternative media in their creation of a space for views which are excluded from the mainstream media, and in their horizontal structure. But although I've used the word 'alternative' throughout these last two chapters, I do have a major problem with the word. It's been so thoroughly co-opted by corporate interests that its usefulness and meaning are in real doubt.

A point made by many writers on different aspects of 'alternative culture' is that it gets commodified and branded. In his book on zines, for instance, Stephen Duncombe notes that 'the underground is filled with people who have heard the Beatles' song "Revolution" and Gil Scott-Heron's "The Revolution Will Not Be Televised" used to sell Nike shoes'.[56] In his book on the Australian 'alternative' rock scene of the 1990s, Craig Mathieson shows how it was bankrolled and created by the established music industry from the beginning – less grassroots than astroturf.[57] And in *No Logo*, Naomi Klein explores how even the identity politics of cultural, ethnic and sexual diversity have been bought up and marketed by corporations; how every struggle for representation ends up as a Benetton ad.[58]

For these reasons, Geert Lovink argues that being 'alternative' has become a meaningless gesture: 'Everyone has to do it,' he says, 'it's required in all management courses. Alternative has been effectively reduced to style. In the media context, this means that we can no longer sell a certain forum – website, radio station, zine – as subversive or even revolutionary. It will have the immediate danger of being turned into a fashion, a lifestyle item.'

In the next chapter we'll look at tactical media and culture jamming campaigns as examples of media activism that try to subvert this trap. These illustrate what Lovink means when he says that 'serious opposition these days has to be on the run'.

CHAPTER 5

Turning Signs Into Question Marks

21 May 1999: Texas Governor George W. Bush is running for President. But a website in his name (www.gwbush.com) isn't doing his campaign any favours. It looks identical to Bush's official site (which is www.georgewbush.com) – the same layout and design, the same pictures. But under the slogan 'Hypocrisy with Bravado', the text highlights Bush's refusal to deny having used illegal drugs, despite his harsh policies on others doing so. 'What, in your own words,' asks the site, 'differentiates George W. Bush's early use of cocaine and that of the felons who are routinely locked away for the same offense, sometimes for years and years, or even forever?' Exploring the site, visitors find invitations to take part in cultural and corporate sabotage projects – edit rented videotapes; jump the fence at Disneyland and request political asylum; insert plastic slaughtered-cow toys into Happy Meals. Asked about the site at a press conference, Bush declares 'there ought to be limits to freedom!'

Quite a sentiment for an aspiring President to come out with so early in a campaign. But, ironically, the people behind the gwbush.com project would be the first to agree with Bush. Except that their activities focus on creating limits to *corporate* freedom, not the freedom of individuals to create websites. In this chapter we'll look at projects such as gwbush.com as examples of *tactical media*. We'll look at how some tactical media use can be seen as an integral part of 'globalisation', rather than as a response to it. We'll then go on to look at some other campaigns which illustrate *culture jamming*, which can be thought of as a subset of tactical media approaches.

The Bush satire site was initially a collaboration between Boston computer consultant Zack Exley and corporate/cultural sabotage specialists ®™ark (pronounced 'art mark').[1] Exley had registered the domain name in 1998 in the hope of perhaps selling it to the Bush campaign. But as he learned more about the alleged tensions between Bush's drug stance and his colourful past, he began to see different possibilities.

'In the beginning', says Exley, 'I wanted to do a copy of the Bush site. Back then I had no reason to think anyone would ever hear about or visit the site. But I thought it would be funny if the Bush people finally stumbled upon the site and found an exact copy – maybe with a few minor and unsettling changes. I had been to ®™ark's site, and seen their copies of corporate websites and figured they had a program that copied them automatically.[2] I emailed them and that was indeed the case. ®™ark made a copy of Bush's front page and layered a reformatted version of the ®™ark website below the top level of the site.'

®™ark

The ®™ark website is an online centre for sabotage funding. Its members co-ordinate a range of cultural activities, from Phone In Sick Day to Corporate Poetry contests.[3] One of ®™ark's aims is to draw attention to the system of corporate power, rather than to the activities of any particular corporation. Another is to reclaim language from corporate appropriation – a point I'll come back to later on. ®™ark are perhaps still best known for their involvement with the Barbie Liberation Organisation, a 1993 event in which they gave US$8000 in funding to a group that switched the voice boxes of some 300 Barbie and GI Joe dolls. The dolls were then returned to toy shop shelves – buyers found that Barbie would bark 'vengeance is mine' on demand.[4]

Other completed projects have included funding a computer programmer to add homoerotic content to the *Simcopter* video game, with around 80,000 doctored copies being distributed; the anti-copyright *Deconstructing Beck* CD, featuring songs constructed entirely from samples of Beck recordings (themselves heavily sample-driven); and a series of earlier clone sites similar to gwbush.com, as Exley mentions above. In these projects, the ®™ark site would be disguised as the official home page of McDonald's or Shell, with the sabotage content layered beneath a copy of the target corporation's site.

While ®™ark have a history which stretches back almost a decade, spokesperson Frank Guerrero stresses the impact that their website, launched in 1997, has had on public responses to their projects.

'Responses are varied, as one might expect in such a volatile marketplace,' says Guerrero in the pitch-perfect corporate tone that ®™ark adopt in everything they do. 'Varied, but increasing tremendously – we have a thousand visitors a day to the website, and many are participating in new discussion groups, which is helping greatly to get the projects worked out with more efficiency. The web allows for greater flexibility of the system and, with the web's popularity, many more viewers and participants.'

A key aim of ®™ark projects is to call attention to the US legal convention of corporate personhood. American corporations have enjoyed the same rights as individual citizens since the *Santa Clara County v. Southern Pacific Railroad* case in 1886. The US Supreme Court ruled in that case that corporations were protected under the 14th Amendment, which had been written to protect the rights of freed slaves, not of economic mechanisms. But while corporations in the US have the same constitutional rights as individual citizens, their limited liability means they don't have the same responsibilities. So ®™ark take advantage of the same loophole, using their limited liability as a legally registered corporation to enable (non-life-threatening) commercial sabotage, as well as drily observing that 'one of ®™ark's ultimate aims is to eliminate the principle of limited liability'.[5] In the case of George W. Bush, the hook was his much-touted campaign 'war chest', and what ®™ark see as the corrosive influence of such corporate finance – and lobbying – on the political process.

The ®™ark site contains a lengthy list of as-yet-unfulfilled sabotage projects, and visitors are encouraged to invest in their realisation on the promise of 'cultural dividends'. Entertainment value aside, these projects do raise crucial questions about corporate rights and sovereignty. Should corporations enjoy the right to privacy, for example, or should their activities be more transparent? There's also the occasional cash incentive – US$2000, for instance, to the first US court which imprisons a corporation under a 'three strikes, you're out' law, or which sentences a corporation to death. Another proposed project argues that 'since US corporations are by law US citizens, it should be possible to marry one'. Marry an actual corporation and you can claim US$200, as well as whatever you can arrange in the pre-nup.[6]

Would you actually see any money? There's a sense in which it doesn't matter. In the next chapter we'll look at the Electronic Disturbance Theater (EDT) and their software, FloodNet, which enables automated virtual sit-ins of websites. The EDT's Ricardo Dominguez points out that the actual funding ®™ark offer is less important than their information network and their ability to draw media attention. The EDT donated FloodNet as an ®™ark project, not for cash, but for a couple of beers.

'In September 1998, at the Ars Electronica festival,' Dominguez says, 'the ®™ark folks came over and said that they would like to buy FloodNet. Of course I knew that meant that they would distribute it via their massive list, which I saw as a great benefit. The real value of ®™ark is the massive network of propaganda that they have, which in itself is extremely valuable. The aspect of exchanging certain cultural values and having this kind of symbolic exchange – the two beers – bespeaks a kind of cultural media activist trend of intimate contact and yet exchanging information on a global level.'

14 April 1999: Bush's lawyer, Benjamin Ginsberg, sends gwbush.com a cease-and-desist letter. It protests about the use of images owned by the Bush campaign and claims that gwbush.com contains links to sites 'that promote violence and degrade women'. ®™ark are later to claim they are deluged with complaints from people who are unable to find this non-existent porn content. Frank Guerrero comments that the group 'do hope to get more cease-and-desist letters in the future'. A few weeks later, on 3 May, Bush makes an official complaint to the Federal Elections Commission (FEC), arguing that the ®™ark site is an unregistered political campaign. And, in a bizarre move, the Bush campaign tries to prevent anyone else from registering Bush-related web addresses by going on a domain name buying spree. Dozens of Bush URLs are set up as aliases to the candidate's site – as late as August 2001, typing in www.bushsucks.com *or* www.bushblows.org *would still redirect surfers to Bush's own official campaign home page.*

'This complaint has big implications,' says Zack Exley. 'It must be viewed in context. There is near unanimous support for Bush among the business community. And being his father's son, he has the complete backing of the military and intelligence communities. And in his complaint to the FEC he spells out his policy on politics on the Internet: all political websites need to register with the government. There's no reason to think his vision of politics on the Internet will not become law when he's President.'

Of course in taking this kind of retaliatory action, the Bush camp were doing just what ®™ark wanted them to do. As with the group's other high-profile actions, the hook for wider media coverage comes from the responses of their targets. Bush's 'limits to freedom' remark generated enormous media interest in gwbush.com – among those to cover it were ABC news, *USA Today* and *Newsweek*, as well as international press from Russia to Brazil. The online version of the *New York Times* featured the story prominently, which Exley credits with contributing to the more than 6 million hits the site claimed in May 1999; the official Bush site reportedly managed around 30,000 in the same period. The publicity also led to a boom in what Frank Guerrero terms 'investor excitement', with similar sites being set up to satirise, among others, Hillary Clinton, Rudy Giuliani and Al Gore. In Australia, Jeff Kennett's unsuccessful campaign to be re-elected Premier of Victoria generated a gwbush.com-style parody site of its own, though one which struggled to be as funny as the site it was subverting.[7]

Gwbush.com was arguably more engaged with actual issues – corporate sovereignty, the drug war – than was Bush's own brochureware HQ.[8] Typically, Bush's campaign site was entirely top-down, offering little more than speeches and bullet-point policy sketches, along with an extensive bio of the candidate from which we learned that his cats are named Cowboy, India and Ernie. Even volunteering was preprocessed, with a list of boxes to be ticked and sent off electronically (sample: 'I'll help by putting a bumper sticker on my car/truck'). A further irony lay in the Bush site's enthusiasm for the wealth-generation potential of information technologies. The new IT economy is going to change the world, but remember – 'there ought to be limits to freedom'.

20 November 1999: As the World Trade Organisation prepares to meet in Seattle, its website (www.wto.org) finds some competition from a site called the 'World Trade Organization/GATT Home Page' (www.gatt.org).[9] The gatt.org page offers quotes attributed to WTO Director-General Mike Moore. He says, for instance, that open trade 'leads to higher living standards for working families, which in turn leads to a cleaner environment'. The site also offers helpful commentary – 'This must be because working families are really dirty, and if you give them a little more money they clean themselves up and stop polluting everything'. As with gwbush.com, gatt.org is an ®™ark clone site, and its commentaries don't really come from the WTO. And as with Bush's 'limits to freedom' intervention, the WTO can't stop themselves from responding. In a WTO press release on 23 November Mike Moore deplores the ®™ark site, saying 'It's illegal and it's unfair'.[10] Inevitably, by responding to the site in public, the WTO simply generates extra publicity and extra traffic, and the story is picked up by CNN,

Wired, Forbes, *and news organisations in Germany and France, among many others.*

®™ark were quick to note the irony of the word 'unfair', as though they should have been playing by some set of rules. Spokesperson Ray Thomas commented that the group was 'supremely flattered that the WTO is acting as if ®™ark were its equal'.[11] The reality is that the gatt.org site was a tactical gesture, a table-turning instance of drawing attention to power and its concealing language. What it shows is that ®™ark are best understood in the context of tactical media – creative and/or subversive uses of communications technologies by those who don't normally get access to them. Tactical media use is a way in for those who feel excluded from their broader cultural environment. And it's a way of voicing grievances about what makes up that cultural, and media, environment.[12]

Tactical media has been both theorised and put into practice at the series of Next 5 Minutes conference/events held every three years in the Netherlands. The program for the most recent of these, in 1999, offers this definition:

> The term 'tactical media' refers to a critical usage and theorisation of media practices that draw on all forms of old and new, both lucid and sophisticated media for achieving a variety of specific non-commercial goals and pushing all kinds of potentially subversive political issues.[13]

Tactical media are different from our earlier examples of *alternative* media in important ways. Media tacticians don't try to consolidate themselves as an alternative – they don't try to create a 'better' radio station or paper, or to establish themselves as, say, a 'Chinese CNN'. Instead, tactical media is about mobility and flexibility, about diverse responses to changing contexts. It's about hit-and-run guerilla media campaigns – 'hit and run, draw and withdraw, code and delete', as two tactical theorists have it.[14] It's about working with, and working out, new and changing coalitions. And it's about bringing theory into practice and practice into theory. Critical Art Ensemble (CAE), for instance, who mix art practice with theoretical critique and tactical actions, illustrate this flexible approach: 'CAE,' they write, 'can be doing a web project one week, a stage performance at a festival the next, a guerilla action the next, a museum installation after that, followed by a book or journal project. Due to collective strength, CAE is prepared for any cultural opportunity.'[15]

'The term "tactical media",' says Geert Lovink, 'was proposed in 1992. It emphasises the use of new technologies, [and] temporary coalitions between artists, designers, activists, theorists and critics who are working both inside and outside the mainstream media.'

®™ark's rolling agenda of projects offers an excellent example of this approach. Different actions and campaigns use whichever media are most appropriate at any given time for any given purpose. An event might call for making a documentary, making a website, making an A4 newsletter, or making a phone call. In making a copy of George Bush's official site and layering their own content underneath it, for example, as well as through using a legally registered domain name, ®™ark and Exley illustrated the tactical approach of exploiting the small cracks that appear in the mediascape through the rapid evolution of technology and the catch-up process of regulatory policy.

In labelling this kind of activity as 'tactical', its theorists are drawing on the work of French thinker Michel de Certeau.[16] In *The Practice of Everyday Life*, de Certeau argued that the study of cultural production and consumption needed to be expanded to examine the *uses* to which cultural products are put by their consumers; not just a study of images or a study of consumer behaviour, but a study of how consumers create meanings and appropriate meanings into their own lives. In relation to TV, for instance, he wrote:

> Thus, once the images broadcast by television and the time spent in front of the TV set have been analyzed, it remains to be asked what the consumer *makes* of these images and during these hours. The

> thousands of people who buy a health magazine, the customers in a supermarket, the practitioners of urban space, the consumers of newspaper stories and legends – what do they make of what they 'absorb', receive and pay for? What do they do with it?[17]

To this end, de Certeau draws an important distinction between *strategies* and *tactics*. A strategy is about exploiting *place* – a business, for instance, that defines its territory and then uses this as the basis for its relations with its customers, works from the privileging of place over time. It's about claiming turf and expanding it, and about using this to create and shape relations with others. McSpotlight, for example, is a long-term strategic activist hub – part library, part portal, it's a *permanent* autonomous zone. A tactic, on the other hand, exploits *time* – the moments of opportunity and possibility made possible as cracks appear in the evolution of strategic place. It's about moments of resistance:

> A tactic insinuates itself into the other's place, fragmentarily, without taking it over in its entirety, without being able to keep it at a distance. It has at its disposal no base where it can capitalise on its advantages, prepare its expansions, and secure independence with respect to circumstances . . . Whatever it wins, it does not keep. It must constantly manipulate events in order to turn them into 'opportunities'. The weak must continually turn to their own ends forces alien to them.[18]

'Tactical media', like 'alternative media', is best seen as a set of options rather than as a monolithic approach. It's a way of thinking as well as a way of doing; a range of tendencies and potentials. Tactical media use can complement alternative media approaches – many Independent Media Centre websites, for instance, are set up as long-term strategic projects; others appear as short-term tactical sites, part of specific events such as the Olympics or May Day (although these too often continue operating, consolidating strategic place).

Many of our earlier examples can be seen as tactical media use – the Panamanian faxes and Iranian cassette tapes mentioned in chapter 3, for example, and the crisis Netcasting of B92. While B92's long-term strategy was to consolidate a space within the media landscape, its use of the Net to subvert Milosevic's censorship is a classic example of the tactical exploitation of time, of moments of opportunity and manipulation. B92's Veran Matic, as we've seen, says that the station is now a de facto public broadcaster, but his account of its early years acknowledges its tactical media elements.

'The program concept of B92, till the late 1990s,' says Matic, 'has been connected with the specific guerilla way of functioning of the station that didn't have an appropriate broadcasting licence, not even a decent transmitter, and thus was often forced to use other media for various campaigns and activities, including street actions and performances as well as Internet activities.'

The history of B92 and the Net illustrates de Certeau's characterisation of the tactic:

> It operates in isolated actions, blow by blow. It takes advantage of 'opportunities' and depends on them, being without a base where it could stockpile its winnings, build up its own position, and plan raids . . . It must vigilantly make use of the cracks that particular conjunctions open in the surveillance of the proprietary powers . . . It creates surprises in them. It can be where it is least expected. It is a guileful ruse.[19]

Think not only of B92's Net use here, but also of tactical actions such as the day they pretended to have been banned in order to mobilise support.

'If History IS "Time", as it claims to be,' writes Hakim Bey in *T.A.Z*, a key tactical media text, 'then the uprising is a moment that springs up and out of Time, violates the "law" of History.' What Bey calls a Temporary Autonomous Zone, like a tactical media strike, is about seizing moments of unexpected opportunity: 'If the State IS History, as it claims to be, then the insurrection is the forbidden moment.'[20]

If the story of new media technologies is being written as one of *convergence*, it's also one of *divergence*. It's worth noting that tactical media is not just the domain of the Internet. With low-cost, high-quality consumer electronics, tactical media use by social movements is infinitely more possible now than it was when groups needed access to a TV station or a printing press. There are camcorders, samplers and digital recording equipment, desktop publishing facilities and photocopiers.[21] There are low-cost, low-power microradio broadcast technologies.[22] Simple mobile phones were used to co-ordinate stunning simultaneous protests across several continents by Kurdish groups in February 1999.[23] As new tools get out there, so activists find moments to turn those tools' uses against power. Moments when the street finds its own uses for things: when South African delegates arrived in London to seek investment during the apartheid era, to be met with a giant swastika projected onto South Africa House;[24] when Sydney activists projected a 15-metre high image of a First Fleet ship above the words 'Boat People' onto the Opera House to protest the mistreatment of asylum seekers;[25] when those excluded from the World Economic Forum in Davos in January 2001 used a large laser to project messages to the world leaders who were attending onto a nearby mountainside.[26] When shoppers take camcorders into department stores in Toronto each Christmas Eve and film the surveillance cameras in the hope of provoking a response from the security teams.[27] When an exile from the former Yugoslavia set up the virtual nation of Cyber Yugoslavia online and made citizenship open to all.[28] And when the camcorder guerillas of *Undercurrents* invaded Shell's London offices in January 1999 and uploaded a live streaming video news report using a digital camera, a laptop and a mobile phone.[29] Each of these was a tactical media strike.

Effective tactical media use is possible even without access to media technologies. All that's necessary is to stage a media event that will draw others' cameras. The Greenham Common women who broke into a military base and danced on top of a half-built Cruise missile silo on New Year's Day 1983 were making a tactical use of the act's image potential.[30] Another example would be the Tactical Air Force of Mexico's Zapatista social movement, which launched a raid on the federal barracks of the Mexican Army in early 2000. The Zapatistas don't *have* an air force – their air raid consisted of throwing hundreds of paper aeroplanes over the fence. The resulting images of confused troops pointing rifles at paper planes bearing messages of peace and solidarity were a tactical media strike.[31]

'A tactic,' writes de Certeau, 'is an art of the weak.'[32] And these examples of tactical media may all be gestures of the weak against the strong. But the example of gwbush.com shows that such gestures can be powerful. That said, the same example also illustrates the fact that the real impact of *tactical* media often comes through hijacking the agenda of the *mainstream* media.

'The reason the media will cover this story,' says Zack Exley, referring to gwbush.com, 'when it won't cover others, is that it involved a personal scandal. The media here is addicted to these kinds of stories. They cannot turn them down. The reason Bush should answer the coke charge is because hundreds of thousands of people are in jail for that type of crime. The media needs an intelligent-sounding rationale for obsessing about personal scandals, and so they will report that one.'

In the same way that alternative media organisations such as B92 can have an impact out of all proportion to their size, media tacticians can leverage their causes onto the public agenda on a digital shoestring: gwbush.com reached more readers than many of the major US political magazines, yet Exley estimates that the site cost just US$210 to create.

'A million people have come to my website and learned about something the media has refused to cover,' he says. 'And they've laughed, because the whole thing is presented as political satire. Since it's funny, they pass it along to friends, and the press has been more interested in it than if I was just a lefty crank yelling about injustice. Without the Net, the only way I could have done this would have been to start my own magazine. The web lets people without millions of dollars become publishers, taking power away from the corporate media.'

Exley points to something important when he mentions the humour of gwbush.com. John Ralston Saul writes that, in a technocracy, 'Comedy remains one of the last weapons we have'.[33] He stresses the importance of satire in the canonical literary tradition – Swift, Voltaire, Cervantes, Rushdie. Why does humour matter? Because, as Saul writes, it is 'The least controllable use of language and therefore the most threatening to people in power'.[34]

Tactical theorist and community organiser Saul Alinsky also stressed the importance of humour and satire in activist campaigns. 'The most potent weapons known to mankind,' writes Alinsky, 'are satire and ridicule':[35]

> It should be remembered that you can threaten the enemy and get away with it. You can insult and annoy him, but the one thing that is unforgivable and that is certain to get him to react is to laugh at him. This causes an irrational anger.[36]

The responses to ®™ark projects from both Bush and the WTO bear this out. But injecting humour into a campaign is also crucial in drawing attention, and in sustaining that interest. I find I look forward to hearing about a new ®™ark project the way I used to look forward to a new record from my favourite band. '*A good tactic is one that your people enjoy*,' Alinsky writes. 'If your people are not having a ball doing it, there is something very wrong with the tactic' [emphasis in original].[37]

'When McSpotlight was being made,' recalls its co-founder, Jessy, 'it was incredibly exciting, and everybody involved with it was obsessed with it because it was the best fun they'd ever had in their lives. It was also probably the most effect they'd ever had in their lives. For the amount of work that was involved (although a fucking lot of work *was* involved!) and the amount of people involved, it had such a large effect that it was so invigorating and so exciting, and that was obvious for the world to see, and that's why it got so much attention. I think it's very important for the people involved in doing a campaign – *it's got to be exciting for them.* If you have a spirit where for everybody involved in it it's the best time of their lives (I don't know how you make that spirit!), then this trickles out and it works.'

Turning up the noise

®™ark don't assume that there is a vantage point from which we can somehow see the corporate media environment from outside. ®™ark are *in* it. They don't just criticise corporations – they *are* a corporation. And they don't just sneer at corporate language – they appropriate it, all the better to subvert it. 'It is impossible to get rid of a world,' wrote Situationist Mustapha Khayati, 'without getting rid of the language that conceals and protects it.'[38]

In his essay on culture jamming (discussed below), cultural critic Mark Dery draws attention to this process of subverting the language which conceals and protects corporate capitalism, arguing that:

> The jammer insists on *choice*: not the dizzying proliferation of consumer options, in which a polyphony of brand names conceals the essential monophony of the advertiser's song, but a true plurality, in which the univocal world promulgated by corporate media yields to a multivocal, polyvalent one.[39]

The emphasis on *choice* is important. Since at least the Thatcher/Reagan years, the term 'choice' has been a vehicle for populist market boosterism and a sustained ideological assault on the concept of public ownership – and on the concept of the public. Citizens have been redefined as consumers, and 'choice' has been elevated as the ultimate value (although the act of consumption itself is taken for granted – 'choice' is assumed to be between products or brands; the option of choosing not to consume is not part of the equation). Direct involvement in the political process means shopping – as the website in chapter 2, whose 'Online Activism' button took us straight to its gift shop, makes clear.

'Choice', then, is a key example of the corporate globospeak that ®™ark subvert by appropriating it for their sabotage projects. ®™ark's online promotional video offers many examples of such subversion and appropriation of corporate language: '... ®™ark is a market-driven system ...', 'the industry leader in bringing sabotage and blacklisted cultural production into the public marketplace ...', 'the cultural bear markets of the 1980s ...'.[40]

'We use this language because it is so effective,' says Frank Guerrero. 'We think that by adopting the language, mannerisms, legal rights and cultural customs of corporations we are able to engage them on their own terms, and also perhaps to reveal something about how downright absurd it can get. Like the words "mutual fund", which to most people means a certain thing having to do with the stock market – by using it in the mix with a slightly different meaning, we may be able to help reclaim it.'

This isn't some kind of attack on globalisation. It's a *part* of globalisation. Resistance to corporate power is as inevitable a part of globalisation as corporate power itself. The widespread friction, hostility, fear and opposition to 'globalisation' are themselves part of the processes the word tries to map. As finance flows across borders and time zones, so too do media images, technologies and people. And those flows are out of sync – the fundamental truth about the relationships between these global flows is, anthropologist Arjun Appadurai argues, *disjuncture*.[41]

As well as the four types of global flow mentioned above (finance, images, technologies, people), Appadurai offers a fifth – the global flow of ideas and ideologies; or rather, of *key words* from ideologies, whose iteration around the globe is also disjunctive, as they are placed in new contexts and taken up by competing groups with competing interests. As key words such as 'democracy', 'freedom' and 'welfare' circulate, their meanings become sites of contestation and struggle. Writing about the Tiananmen Square events, for example, McKenzie Wark makes the point that the label of 'democracy movement', as applied by many in the Western media, is a problematic one. 'Democracy', in this analysis, was a kind of free-floating term, invoked to serve competing interests. Wark quotes Geremie Barme and Linda Jaivin's comment that:

> democracy was not one of the movement's strong points. Rather, its overriding theme was one of protest – against dictatorial one-party rule, a lack of both individual and group autonomy, economic and political mismanagement, and government unresponsiveness to its people's concerns.[42]

Appadurai restricts his discussion to key words and concepts from Enlightenment-derived democracy – terms such as 'freedom', 'rights', 'representation' and 'democracy' itself.[43] But Dery's emphasis on 'choice' opens the way for us to see that this is also a useful concept in understanding the global diffusion of core concepts and language repertoires from other ideologies – such as the discourses of corporate consumerism.

®™ark, most obviously, appropriate the key words of corporate consumerism and the rhetoric of 'choice', 'free markets', 'dividends' and 'opportunity'. What's equally interesting, though, is the ways in which corporations can counter this by, for example, turning to the key words, rhetoric and values of environmentalism. Just as oppositional groups have learned to speak the corporate language of 'choice', so corporations have learned to exploit the language of 'green culture'. These competing rhetorics of 'choice' and 'sustainability' are as much a part of globalisation as a new Nike factory. Consider 'life sciences' company Monsanto, for instance, which became the focus of global opposition to the use and patenting of genetically modified organisms, and which developed the 'Terminator' technology to render seeds sterile after a single use, forcing farmers into an annual dependence on the company's products.

Monsanto's website talks of the company's 'family of 30,000 employees', who are committed to 'environmental sustainability'. An online slide show develops these themes of sustainability and family, beginning with images of the Earth seen from space alongside text which declares 'There's a family that lives here. A family that's lived here for thousands of years, getting to know the land and the oceans and the sky above.'[44] No mention is made of patenting them.

Just say no

19 November 2001: CNN Headline News cuts to an ad break. Viewers who don't immediately change channels see a burping cartoon pig superimposed on a map of the US. 'The average North American,' says a voiceover, 'consumes five times more than a Mexican, ten times more than a Chinese person, and thirty times more than a person from India. We are the most voracious consumers in the world... a world that could die because of the way we North Americans live. Give it a rest! November 23 is Buy Nothing Day.'

Buy Nothing Day is co-ordinated by the Vancouver-based Media Foundation, generally known as Adbusters.[45] Adbusters was founded in 1989 by Kalle Lasn after he had tried unsuccessfully to buy airtime for an advocacy message opposing a Canadian logging industry TV advertising campaign. Although this 'uncommercial' or 'subvertisement' was never shown, the resulting media furore contributed to the withdrawal of the logging industry media campaign, and the first issue of the quarterly *Adbusters* magazine was published later the same year. Their website includes a gallery of very professional parodies of well-known ads, such as their take on Calvin Klein's 'Obsession' campaign, which features a bulimic model snuggling up to a toilet bowl.[46] The important thing to note about this subvertisement is that it is not an attack on a particular perfume campaign, but on advertising itself.

Like ®™ark, Adbusters have a history which predates the web. But again like ®™ark, they credit the web with providing them with a platform from which to take their campaigns and projects to a wider, international audience. Adbusters' most successful use of the web is in making multimedia resources available to participants for local actions in support of global campaigns. For Buy Nothing Day, for instance, they offer resources including a downloadable Quicktime version of the TV 'uncommercial'; Christmas Gift

Exemption Vouchers, which pledge 'quality time' instead of presents; and posters which can be printed out directly from browser windows.

'At first we didn't have too much faith in the website,' says Lasn. 'It just felt like a sink-hole for money. We kept on throwing bucks at it and coming up with stuff and it didn't seem to do anything. Then something really phenomenal happened. We had this annual Buy Nothing Day, and it was growing into quite a large campaign here in North America. But then in 1995 we put this campaign onto our site. We allowed people anywhere around the world to download black-and-white posters, and we put up Quicktime versions of some of the TV spots. And suddenly that year that campaign took off worldwide in a way that absolutely dumbfounded us. Then we suddenly woke up to the power that the Internet gives us in taking a relatively local campaign and turning it into a global blast.'

In 2000, the event was marked from Panama (with a music festival and barter market) to Finland (with a televised spaghetti western-themed uncommercial).[47] In Sydney, students set up a free coffee stall outside a Starbucks and restocked the shelves of a music shop with home-burned CDs of Metallica tracks downloaded from Napster. Buy Nothing Day banners were hung in shopping malls from Philadelphia to Dunedin, New Zealand. The year before, there was a 45-minute dance party in New York's Times Square, and demonstrations, performances and credit card cut-ups were reported from Brazil, South Korea, Israel and Japan.[48] In previous years, the campaign made the front page of the *Wall Street Journal*, saw one-off 'No Shop' shops stocked with empty shelves, and saw activists arrested after hanging a 56 square metre banner at America's largest shopping mall. Malls are the focus for many Buy Nothing Day actions – as strategic sites of consumer culture, they're an obvious place for tactical interventions.

While Adbusters' media use is mainly *tactical*, they also draw upon some of the characteristics of *alternative* media, particularly with their emphasis on the creation of horizontal linkages among the users of their website. The Buy Nothing Day section includes contact details for sympathisers and supporters around the world:

these range from established major groups, such as Greenpeace in Germany, to a high school student in Toronto, Canada, who planned to mark the occasion by encouraging fellow students to bring in a packed lunch rather than eat in the fast food chains at the school cafeteria. But as we'll see below, in the case of Adbusters, the mix of tactical and alternative approaches points to a tension in their work. If tactical media can be seen, on one level, as an attempt to rethink the strategic alternatives of the old Left, Adbusters are still very attached to those older approaches: this manifests itself in an at times contradictory approach to their audiences (we'll come back to this point later on).

Like many of the groups in this book, Adbusters explicitly identify the media as adversaries. But where some groups try to draw legitimacy from the media while simultaneously exploiting their visitors' perceived mistrust of the corporate mediascape, Adbusters' approach is a far more explicit attempt to monkeywrench the media machine. They deploy their full range of cultural sabotage and interventionist techniques, which they label *culture jamming*.

Adbusters style themselves 'culture jammers headquarters'. Culture jamming is an attractive label, one which describes a range of activities which would fall under the broader heading of tactical media, from altering billboards to street events and pranks. Culture jammers deploy the techniques now commonly identified with a postmodern aesthetic: appropriation, collage, ironic inversion and juxtaposition. And they also, as we'll see below, owe a heavy debt to the ideas of the Situationist International, the radical movement of the 1950s and 1960s which centred around Guy Debord.

The term 'culture jamming' is generally credited to the rock band Negativland. In Craig Baldwin's documentary, *Sonic Outlaws*, a member of the band defines it as simply 'going in where you're not supposed to be on the airwaves and screwing everything up'.[49] It came into wider use through a pamphlet by Mark Dery. For Dery, jamming is 'guerilla semiotics', the shape of 'an engaged politics... in an empire of signs'.[50] As with Downing's emphasis on the long history of alternative media, Dery situates jamming in a timeline which includes:

> Russian samizdat . . . Situationist detournement . . . the underground journalism of '60s radicals . . . parody religions . . . workplace sabotage . . . the ecopolitical monkeywrenching of Earth First! . . . the insurgent use of the 'cut-up' collage technique proposed by William Burroughs and others.[51]

Resistant media, write Critical Art Ensemble, should call attention to the construction of signs, or else cause confusion so that simulated media constructs can't work: 'Confusion should be seen as an acceptable aesthetic. The moment of confusion is the pre-condition for the scepticism necessary for radical thought to emerge.'[52] Culture jamming works through such use of confusion. But jamming is also self-reflexive media activism – jammers use the media to draw attention to issues and problems with those same media. What makes jamming more than just juvenile trespassing is its *media literacy* emphasis.[53] Culture jamming turns familiar signs into question marks.

'Culture jamming is a way to fight back against advanced consumer capitalism,' says Lasn. 'I see the kind of consumer culture that we have built up over the last many years as being unsustainable. It's a culture that drives the global economy in a way that will eventually make it hit the wall. Culture jamming is a way to get this dysfunctional culture to bite its own tail.'

'First of all we think the world must be changed'

These were the opening words of the founding document of the Situationist International in 1957.[54] Read today, this document, like other Situationist writings, is a bizarre mix of the banal and the insightful, the dated and the prescient, the hopelessly obscure and the shockingly lucid. Adbusters, like other culture jammers, owe a debt to the Situationists, who provided the theoretical background for the near-revolution in Paris in 1968.[55] The welcome page of the

Adbusters website was for a time a Situationist slogan – 'live without dead time', one of the most frequently sighted graffiti of May 1968, when *France-Soir* wrote that the students were 'fighting advertising on its own terrain with its own weapons'.[56] Fighting advertising with its own weapons is central to any attempt at undermining what the Situationists called the *spectacle*: the integrated, commercialised cultural space in which 'Everything that was directly lived has moved away into a representation.'[57] Culture jammers act on what the Situationists theorised, emphasising the centrality of advertising to the spectacle.

The Situationist influence goes beyond the recycling of 30-year-old slogans: it can be seen in Adbusters' own language and in their goals. Cultural critic Greil Marcus describes the Situationists' voice, in examples such as 'boredom is always counterrevolutionary', as 'a blindside paradox of dead rhetoric and ordinary language floated just this side of non sequitur, the declarative statement turning into a question as you heard it: what does this mean?'[58] But the response can go further than 'what does this mean?' – even to acknowledge a Situationist question such as 'what are the politics of boredom?' means engaging with the premise that boredom might be political, and that 'free' time has been commodified as 'leisure' time. The influence of this style of writing is clear in Adbusters' slogans such as 'participate by not participating' or 'it's time to unsell the product'. Such lines aim to make us do a double-take, to arrest our attention (as advertisements would), but also to challenge the premises which lie behind them (as advertisements would not).

The Situationist influence is also at the heart of one objective behind the culture jammers' project – to encourage acts of negation that become affirmations; to bite the hand that spoonfeeds; to say *no* (to shopping, to the news, to work) in ways that can lead to an unexpected *yes*. Drawing a line from the Situationists to punk, in writing about the Slits, Marcus describes 'four teenagers who hadn't the slightest idea of how to do anything but climb onto a stage and shout. They said "Fuck you"; it meant "Why not." It was the sound,

Jon Savage wrote long after the fact, of people discovering their own power.'[59]

A key tactic here is the practice of *detournement* – ripping an image out of its original context and setting it in a new one, creating a synthesis that calls attention to both the original context and the new result.[60] Think, for instance, of ®™ark's appropriation of corporate language. Situationists Guy Debord and Gil Wolman explain *detournement* in this way:

> When two objects are brought together, no matter how far apart their original contexts may be, a relationship is always formed... The mutual interference of two worlds of feeling, or the bringing together of two independent expressions, supersedes the original elements and produces a synthetic organization of greater efficacy. Anything can be used.[61]

Umberto Eco argues that the most complex systems are correspondingly vulnerable to the effects of random variation: 'All you have to do is insert into its circuit, perhaps by telephone, a piece of "wild" information.' A kind of *detournement*, in other words, or what Exley and ®™ark did to the Bush campaign, exposing a crack in the tightly wrapped surface of media management.[62]

Everything you know is wrong

One sophisticated application of this might be Richard Metzger's Disinformation site, which tries to call the whole status of information, and its means of delivery, into question. Disinformation is an all-out assault on the journalistic conceit of objectivity. It's a gateway to the paranoid, the subversive, the conspiratorial, and the plain loopy. *X-Files* fans will probably have realised by now that the truth was never really out there. But it's probably in here. Disinformation mixes essays and links on the political and the paranoid – there are articles about the World Trade Organisation and about killer bees; about East Timor and about the ever-popular 'Bush Nazi Coke Moonie Connection'.[63]

'We call "reality" (not just the media or business's version of reality) into question at the root level,' says Richard Metzger, 'by including UFO stories and *X-Files* type materials with "left" politics (or whatever our brand can be described as).' To navigate its sprawling content, Disinformation also features a search engine, but one with a highly subjective and targeted database – type in something harmless like 'cats', and you're trapdoored into a parallel universe of pet worship, killer felines and Mat Cat Disease. Like the Natural Selection search engine, discussed in the previous chapter, this is more than just a novelty item. Just as Disinformation's content tries to undermine the conceit of journalistic objectivity, so its search engine works to undermine the authority of regular engines, by calling attention to their in-built assumptions.

What's interesting about Disinformation's blend of activism and paranoia is that it assumes from the start an active, questioning, sceptical audience – an audience who not only don't need but don't accept the notion of journalistic objectivity. Aspiring to become the 'anti-CNN', Metzger says he sees the site as more than just entertainment – he sees it as essentially an activist project.

'Conspiracy theory is like a ride through a carnival funhouse,' says Metzger. 'A hall of mirrors. We want to fuck with people's heads, but where it deposits them afterwards is out of our control, obviously. Nothing is objective any more. The so-called "objectivity" of the media is a conceit that the human race can't take into the new millennium, can it?'

Disinformation is about feeding noise into the system – a tactic shared by many tactical media practitioners, Net artists and culture jammers. There's the Digital Landfill, for instance, which challenges the idea of the web as a publishing medium. Visitors can submit text or images and, by skewing the HTML code, it will be re-presented as part of a chaotic collage.[64] Or there's re-m@il, another tactical art project, to which you can forward any of your unwanted email, to be answered on your behalf by other visitors to the site.[65] Or, as we saw in the previous chapter, there's Natural Selection and its creators, Mongrel.[66]

'The idea of an overloaded information system,' says Mongrel's Matthew Fuller, 'that one lives at a particular moment that is more dense with meaning than any other, is a historical conceit. I don't know if the tools exist to determine whether this is so or not, but we don't need to flatter ourselves that we can't cope. There are more interesting ways of fucking up. Mongrel and its collaborators always come into these systems as noise, as an irritant, if not as the cleaners. So getting grit into the parsing comes naturally. It's time to take advantage of it. One of the ways of doing that is by openly multiplying ways of seeing, knowing and doing in the world. Not as some kind of sick pluralist glad-handing, but as part of a sensuous arms race against the one world order.'

Media cartels, Media Carta

Article 19 of the Universal Declaration of Human Rights: 'Everyone has the right to freedom of opinion and expression; this right includes freedom to hold opinions without interference and to seek, receive and impart information and ideas through any media and regardless of frontiers.' In 1998, inspired by this ideal, Adbusters launch an online petition to raise support for their Media Carta campaign. Media Carta argues that what's needed is access for all to the established media – a new diversity and public sphere. Pointing to the sheer scale of the top half-dozen media corporations, Adbusters argue that the key to these corporations' power is their vertical integration, through which the same material is marketed as a book and a film, a TV show and a CD, a video game and a ride. 'Between them,' write Adbusters, 'the media giants have the means to produce a never-ending flow of social spectacles . . . a jolly, terrifyingly homogenized Las Vegas of the mind.' Now a Net petition isn't likely to bring Disney or Viacom to its knees. But the value of Media Carta is in its attempt to raise awareness of the emerging global communications cartel.[67]

The German media giant Bertelsmann becomes the biggest English-language publisher in the world with the purchase of Random House and Transworld, before going on to buy into Napster. In the UK,

Microsoft joins forces with British Telecom in an Internet television venture, and in Australia it teams up with the largest commercial TV network, Channel Nine. Not to be outdone, Nine's nearest rival, the Seven Network, announces a co-operative technology agreement with Intel, and a deal with AOL. Meanwhile, the Australian Broadcasting Corporation strikes a deal with Telstra to provide news content for its website, initially agreeing to regularly discuss with the telco the 'mix and variety' of news and to accept 'reasonable' suggestions from them. And back in the UK, the BBC signs a commercial agreement with the Discovery Channel, part of the media conglomerate TCI. Or – hang on – was that ICI? FBI? REM?

Keeping up with these guys is a full-time job.[68] The AOL/Time–Warner deal may have hogged the headlines, but it was only one example – albeit a particularly dramatic one – of a trend, as media, communications and technology companies stay busy with the real work of mergers, acquisitions and alliances.[69] The Media Carta campaign asks, 'What does freedom of speech mean in this kind of mental environment?'

Every year, in a kind of ritual, Adbusters approach the main American TV networks to buy airtime for the Buy Nothing Day uncommercial. And every year they are rejected. Not, apparently, because the ads are below broadcast quality, but because of their *content*. Various network executives have made this impressively clear: NBC's commercial clearance manager wrote to Adbusters that 'We don't want to take any advertising that's inimical to our legitimate business interests'. A CBS executive trumped this by protesting that the ad was 'in opposition to the current economic policy in the United States'.[70] In 1999, a group of French jammers ran into exactly the same responses when trying to buy airtime for their own Buy Nothing Day spot.

Why is this? How pernicious and corrosive can it really be to recommend that people try going for 24 hours without giving in to the urge to shop? For Lasn the significance of this is *not* that his group in particular has been discriminated against. Rather, in his words, 'whole classes of information about transportation, nutrition, fashion and sustainable consumption' are being kept from the public airwaves because of the influence of corporate sponsors.[71] Lasn hopes eventually to take a case to the World Court in The Hague, under Article 19 of the Universal Declaration of Human Rights, arguing that the airwaves, as a public resource, should be open to non-commercial advocacy messages. He argues that this fight for non-commercial public space is one that more and more people are going to join.

'I think a lot of people feel that the old grand movements of the past are really dead now,' Lasn says. 'So more and more people are waking up [to the fact] that the fight of the future is not about race, it's not about gender, it's not about left and right – it is about culture. At the moment our big enemy is a dysfunctional culture that's controlled by corporations. And I think that's why culture jamming has a chance of being the next big social activist movement.'

As part of the Media Carta campaign, Adbusters' website offers several tactics for participants to take up. These include promotion of the annual TV Turnoff Week, held each April, and also lobbying for regulations which would require TV networks to allocate two minutes each broadcast hour to citizen-produced, non-commercial advocacy messages. They also point to the US government's antitrust suit against Microsoft as an example to activists who aspire to pressuring governments to enact tougher cross-media ownership laws, and publicise the case of WHDH-TV in Boston, which had its licence revoked in the 1960s after residents filed a petition complaining about 'shoddy nightly news broadcasts'. The group concede that no one has since duplicated this success, but the precedent remains.

Lasn describes the Media Carta campaign as 'the great human rights battle of the information age'. But in their insistence that the mainstream media mould their audiences into narcotised suckers – '*a Las Vegas of the mind*' – Adbusters rather undermine their overall project. Can Adbusters' own community really be active, sophisticated, cultural critics – *culture jammers* – and passive stooges of corporations at the same time? This apparent contradiction, as sociologist Manuel Castells notes, is a common one: 'it is precisely those thinkers who advocate social change,' writes Castells, 'who often view people as passive receptacles of ideological manipulation'.[72] The existence of projects such as Adbusters and ®™ark, while maybe gestures of the weak against the strong, is itself a reminder that the media oligopoly is hardly complete – at least for now. And concentrated ownership doesn't necessarily mean that successful brainwashing of audiences will follow – control of the means of production is not the same thing as control of the production of meaning. Lasn acknowledges these contradictions: 'I think that in this crazy postmodern age of ours, we're all stuck in this hall of mirrors,' Lasn says. 'All of us move from being zombied out in front of the TV set one moment, to suddenly trying to be empowered and really understanding what's going on, and trying to fight back against this advanced consumer capitalism that's somehow eating us up in all kinds of ways. I think that's just part of the postmodern condition, and that contradiction is inherent in our age, and in many of our psyches and our personalities. I myself can't quite get away from it: I oscillate between the two states myself and don't quite understand it.'

Coming up

In the next chapter we'll stay with the idea of tactical media, looking in detail at one intercreative example – electronic civil disobedience – and how it relates to longer-term strategic approaches. Electronic civil disobedience is still an emerging area, but a key tactic is the virtual sit-in, which, says one of the activists we'll meet in the next chapter, is about 'creating the unbearable weight of human beings in a digital way'.

CHAPTER 6

Hack Attacks and Electronic Civil Disobedience

21 October 1967: Tens of thousands of anti-Vietnam War demonstrators converge on the Pentagon in Washington DC. Co-organiser Abbie Hoffman has applied for a permit to levitate the Pentagon some 90 metres in the air. If it levitates, Hoffman says, it will be exorcised, evil will be cast out, and the war will end. The Pentagon's General Services Administrator haggles him down to 3 metres and grants the permit. The Pentagon has left nothing to chance, with troops committed in the city for the first time since 1932, and the 82nd Airborne Division on standby just outside the capital. A total of 8500 military personnel are deployed. Just in case.[1]

'In perpetrating a revolution,' wrote the great political theorist Woody Allen, 'there are two requirements: someone or something to revolt against and someone to actually show up and do the revolting'.[2] *Thirty-one years after the Pentagon levitation, in September 1998, a group of four activists called the Electronic Disturbance Theater prove that no one now has to actually show up. Using a piece of software called FloodNet, they too blockade the Pentagon. Only this is a virtual sit-in, a blockade of the Pentagon's website. In the 1960s, it took tens of thousands of people to physically gather at the site to provoke a response. In the 1990s, a group of just four people organise a gathering at the website, and provoke some unexpected responses – not least from the Pentagon itself.*

In this chapter we'll look at the Electronic Disturbance Theater (EDT) and their virtual sit-ins – *electronic civil disobedience* – as a key example of what is becoming known as *hacktivism*. It's a label invented by journalists, not by activists, and what it means is still partly up for grabs. But one thing the 11 September 2001 assault on the Pentagon showed starkly is that hacktivism is *not* terrorism, despite its frequent linking to terrorist activity in mainstream media.[3] We'll look at whether or not it's hacking in any real sense. And we'll consider the risks for activists in allowing others to define their actions for them. In accepting the label 'hacktivism', the EDT are taking on all the baggage of 'hackers'. It's a word that has lost its original meaning – so in using the words 'hack attacks' in the title of this chapter, I'm thinking as much of attacks *on* hackers as *by* them.

The theatre is open

The EDT is a group of four activists: writer and theorist Stefan Wray, digital artists Ricardo Dominguez and Carmin Karasic, and programmer Brett Stalbaum. They attempt to draw attention to the struggles of the Mexican Zapatistas through the practice of what they term 'electronic civil disobedience'. But insofar as the wider public is aware of this, it's as 'hacktivism'. It's an empty term as yet, in that the practices it tries to label are still emerging. One preliminary contribution towards a definition might be that hacktivism is an engaged politics which seeks solutions in software, in the search for a specific technological fix to a social problem. Such electronic civil disobedience, say its pioneer theorists – Critical Art Ensemble – is 'hacking that is done primarily as a form of political resistance rather than as an idiosyncratic activity or as a profit- or prestige-generating process'.[4]

On 25 August 1999, an email from the EDT circulated through lists around the world:

> Once More Dear Sisters and Brothers, We Must Act To End This WAR That Dare Not Speak Its Name. Once More We Must Do Whatever Each of Us Can Do to Stop This Attack on the Zapatista Communities.

> Therefore We Are Calling for a 24 Hour Digital Storm On the Mexican Government. The Storm Will Start At: 12am (Mexico City Time) On August 26th, 1999 and End At: 12am (Mexico City Time) on August 27th, 1999.[5]

This call for a 'digital storm' was an alert for a *virtual sit-in* – it's not a call to action in the real world, such as a blockade of a physical space. It's an event which happens wholly within the networked virtual space between the participants and their target. In a virtual sit-in, supporters swarm a targeted website and deluge it with bogus requests to reload the page. The idea is that the server will be unable to cope with the volume of traffic and that this will block out other visitors.

'Say that you want to go read the *New York Times* online,' explains Ricardo Dominguez. 'But you look at it and say, "this is yesterday's news" – it's loaded in your cache. So you hit the reload button and your machine goes knocking on the server at the *New York Times.* The server opens the door and says, "yes, can I help you?" and the browser says, "I want this page, index.html" and the server gives it to you and it falls freshly into your page. So what FloodNet does is it takes into account, like a hamster wheel, how many hamsters are coming. The more hamsters come, the faster it hits the reload button. So that means more knocking on the server – *'we request this, we request this'* – until the server is disturbed, it slows down. At some point it might even say, "look, I can't even answer your question, because so many people are requesting it". If you can look at the browser as an *agora*, a commons, this particular tool is a way to create a sign that blocks the door. Now you may have a thousand people in front of this door – it's not going to destroy the door and it's not going to destroy the building. But it's going to make it very difficult for you to get through that door. We're not cracking into the system, we're not getting root index control and changing the website or damaging the data. All we are doing is creating the unbearable weight of human beings in a digital way.'

One obvious objection to this is that it takes all the weight out of the activist gesture – people aren't putting their bodies on the line. Instead, a virtual sit-in can be seen as a kind of computer game (a mistake that the EDT website warns new visitors against). But a screen can connect as well as separate. It can enable people to take part who might otherwise never have been able to. The virtual sit-in, like many of the other Net activist tactics in this book, is essentially an electronic update of an established protest gesture – it's about backing into the future. But it's significant in that it takes cyberspace as the actual site of action. So the virtual sit-in also represents a move towards using the technical properties of the new media to formulate new tactics for effecting social change. At the same time, the concept of the virtual sit-in also illustrates one of the *advantages* of backing into the future: if it's a familiar tactic, people can quickly see where you're coming from.

'At first movement,' says Dominguez, 'I think it's much easier for people to manifest themselves if they can consider that it's somehow bound to a history that they know. If I said "virtual sit-in", it had a kind of pedagogical usefulness that the term that I would prefer – "electronic civil disobedience" – did not have. And it's something that they're not afraid of – they understand what a sit-in is, they understand that civil disobedience is about non-violence. It becomes a way in which you can re-embody in cyberspace a certain lived condition which people understand traditionally from Gandhi, from Civil Rights.'

FloodNet was conceived of as a year-long project, throughout 1998, although actions have continued since then. The first action, Dominguez claims, drew 18,000 participants over two hours. Many emailed to report that Mexican President Zedillo's website was inaccessible during this period. Participation in the EDT's virtual sit-ins increased with each action: the first experimental use of FloodNet in April 1998 attracted participants from more than 8000 different Internet addresses, alerted by postings and cross-postings on a wide range of mailing lists. By the 'digital storm' of August 1999, this number had tripled, with participants coming from as far afield as South Africa, Hong Kong and Iceland.

To take part in a virtual sit-in, we gather at a pre-announced website and activate the FloodNet software, which is directed towards a target web page – previous targets have included the White House, the Pentagon, the Frankfurt Stock Exchange and various Mexican government sites. The parallels with a physical sit-in are developed on the FloodNet page itself. To take part in this 'digital storm' we gather in the 'FloodNet Foyer'. We're met with a set of instructions – browser configurations necessary for the software to be engaged and to protect the participant from retaliatory responses – and a set of warnings: 'This is a protest, it is not a game, it may have personal consequences as in any off-line political manifestation on the street.' We're warned that our computers' IP addresses will be collected 'by the government', in the same way that our pictures might be taken during a street action. We're warned of possible damage to our computers, in the same way that 'in a street action the police may come and hurt you'. And we're warned that a large-scale FloodNet action may have effects on bandwidth, slowing or preventing access for others as the network connections are clogged with virtual blockaders. Here again there's a parallel with a real-world sit-in – passers-by may be inconvenienced if their street or office building is blockaded.[6]

FloodNet also makes use of the web browser's 404 file-not-found function. If I request a page which doesn't exist – such as www.nytimes.com/Graham.html – the server of course won't be able to find it, and will return a 404 error message: 'Graham was not found on the *New York Times* website'. 'We reverse this gesture,' says Dominguez, 'into a political condition that reflects the real conditions of politics outside. That is, we requested at President Zedillo's website: "Does justice.html exist on this page? No, justice does not exist on President Zedillo's website." And who sees that? The system administrator. When he's looking at the daily list of what people are looking at, what are they requesting, they're requesting "Justice! Justice! Democracy! Democracy! Zapatistas! Zapatistas!"'

The Zapatistas themselves have an established reputation as one of the first groups to exploit the Internet as part of their communications strategy. It's appropriate, then, that the Zapatista cause should be the inspiration for the EDT, who are among the first to try to shift the actual site of political action into cyberspace itself. Sociologist Manuel Castells describes the Zapatistas as 'the *first informational guerilla movement.* They created a media event in order to diffuse their message, while desperately trying not to be brought into a bloody war' [emphasis in original].[7]

But the Zapatistas can also be seen as the first *culture jamming* guerilla movement, an ingenious *detournement* of the predictable – and hence controllable – imagery and rhetoric of insurgencies. Subcomandante Marcos's ever-present pipe and balaclava, his refusal to accept the label of leader, and the cryptic parables through which he communicates, make it hard for their opponents to frame the Zapatistas on easily understood terms. Marcos's signs work as question marks, and this opens up a space of potential advantage for the Zapatistas – an opportunity to try to set the debate on their terms.

The Zapatista events had begun as a familiar kind of insurgency: the forces of the Ejercito Zapatista de Liberacion Nacional (EZLN) first rose on 1 January 1994, the day in which the North American Free Trade Agreement came into operation, with ugly implications for the corn-growing economy of the Chiapas region. This erosion of local control over the economy carried nasty reverberations from Mexico's colonial past – a point which was made in the opening lines of the EZLN's first declaration: 'We are a product of 500 years of struggle'. The EZLN moved to occupy population centres in the Chiapas region, declaring a list of demands and inviting overseas organisations and media to the area as observers. The Mexican government reacted predictably, with a heavy military and police presence. What was *not* predictable was the way in which the Zapatistas were able to set the media agenda, circumventing the government's attempts at crisis management.[8]

The Net enabled the Zapatistas to spread their bulletins, communiques and alerts without the necessary endorsement of the mainstream media; more importantly, it enabled them to build networks of support and pressure from NGOs and activists around the world. But despite their successes in creating a media event and dictating its terms, the Zapatistas continued to struggle against very non-virtual forces. On 22 December 1997, more than 40 members of a Zapatista community were shot dead at Acteal by state-supported paramilitaries. This massacre provided the impetus for the evolution of FloodNet. Ricardo Dominguez began to look for others who could help him put into practice the idea of electronic civil disobedience.

Electronic civil disobedience

'Electronic civil disobedience' is a term introduced by Critical Art Ensemble (CAE), a five-member group who've been mixing digital art theory and practice since the 1980s – writing four books and exhibiting and performing widely. While CAE didn't begin to think about the Net until the 1990s, their involvement in AIDS activism in Florida in the mid-1980s started them thinking about the importance of developing new tactics.

'ACT UP was significant for CAE for two reasons,' says CAE member Steve Kurtz. 'One, it showed CAE how political and cultural resistance could be integrated: a political wing concerned with policy, and a cultural wing concerned with representation. Two, it showed the limits of street action. It got to the point in New York in the late 1980s where the cops were just letting ACT UP protesters be as disobedient as they wanted. When a demo appeared, they would close off the street, and just wait for them to go home. The protesters were not even worth arresting any more. That is when we knew something was wrong with traditional action.'

By late 1991, CAE had begun to move towards the idea of electronic civil disobedience, drawing on their experiences with ACT UP to argue that the street was losing its power as a site of intervention. At this point they were still thinking about video and television as an alternative to traditional civil disobedience. But in 1993 some CAE members read Bruce Sterling's book *The Hacker Crackdown*, and became interested in hacker exploits. 'We started thinking,' Kurtz says, 'what if this practice was radicalised?'

In 1995 CAE published a collection of essays titled *Electronic Civil Disobedience*. In it they argued that gathering in the streets was a dead tactic. If power was now nomadic and decentralised, then physical protest at a physical site was like picketing a monument to dead capital. Instead, they proposed hacker actions against the cyberspace presence of institutions. 'Right now,' they wrote, 'the finest political activists are children.'[9] By this they meant that teenage hackers were alone in looking for ways to penetrate and subvert the new systems of informational power. The solution to the problem of dead tactics was new alliances between hackers and activists – electronic civil disobedience. As with traditional civil disobedience, their model set trespass and blockades as central tactics. And, to combat decentralised power, they proposed decentralised, cell-based organisation: small groups of from four to ten activists. 'ECD is CD reinvigorated,' they wrote. 'What CD once was, ECD is now.'[10]

Except, as they acknowledged, it was still sci-fi at that point. Hackers were not politically active (partly because serious hacking skills are so time-consuming to acquire and maintain that they leave little room for anything else). CAE wrote that hackers would not become politicised, as their prime motivation was intellectual curiosity: disturbing the system would conflict with their desire to master it. This fits with the views of other commentators around that time. Hackers, writes Sterling in *The Hacker Crackdown*, are not just curious, but are possessed by 'a positive *lust to know*' [emphasis in original].[11] Finding stuff out is an end in itself. 'The problem,' wrote CAE, 'is much the same as politicizing scientists whose research leads to weapons development. It must be asked, How can this class be asked to destabilize or crash its own world?'[12] In his book *Hackers*, Steven Levy also noted that 'hackers do not generally set about to create social change – hackers act like hackers'.[13] And Sterling points out a further problem in politicising hackers: as their behaviour is generally illegal, it's difficult for them to speak out publicly or take a position – drawing attention to themselves *as hackers* is inviting arrest.[14]

Like all concepts of civil disobedience, CAE's new model owed a debt to Thoreau. The legacy of Thoreau's ideas is perhaps best seen as growing from his line 'It is not desirable to cultivate a respect for the law, so much as for the right.'[15] In the EDT's attack on the Pentagon – as in Abbie Hoffman's – Thoreau's influence is in the simple denial of the Pentagon's authority, the refusal of deference. His argument that it's morally right to refuse to obey an unjust law has been a major influence on leaders such as Gandhi and Martin Luther King.[16]

But CAE argued that electronic civil disobedience should be *surreptitious*, in the hacker tradition. This is, of course, a reversal of the tradition of (pre-electronic) civil disobedience. Gandhi, for instance, said that civil disobedience could not be clandestine:

> The lawbreaker breaks the law surreptitiously and tries to avoid the penalty, not so the civil resister. He ever obeys the laws of the State to which he belongs, not out of fear of the sanctions but because he considers them to be good for the welfare of society. But there come occasions, generally rare, when he considers certain laws to be so unjust as to render obedience to them a dishonour. He then openly and civilly breaks them and quietly suffers the penalty for their breach.[17]

CAE, in contrast, see electronic civil disobedience as 'an underground activity that should be kept out of the public/popular sphere (as in the hacker tradition) and the eye of the media'.[18] They propose multiple, diverse actions from cell-based actors, instead of a mass public protest movement. This emphasis on surreptitious, clandestine operations appears at odds with the classic tradition of civil disobedience; it seems counter-intuitive. But Steve Kurtz argues that there are advantages to using the term 'civil disobedience', and that this is justified. 'It's similar enough,' he says. 'Our model uses blockage and inertia tactically, and we believe in peaceful action. Also, "civil disobedience" has good associations. We needed to disassociate ourselves from the conventional view of hacking.'

If the work of CAE is a call to action, then FloodNet is a clear response – although one that CAE aren't happy with.[19] Where CAE proposed hacker activism against the inner systems of cyberspace, FloodNet is instead more of a symbolic attack than a genuine attempt to crash the system. Targeting websites, the visible front-end of global flows of money and power, has more in common with a non-violent occupation, or with culture jamming, than with the monkeywrenching of cracker attacks. With FloodNet, the EDT moved electronic civil disobedience away from the clandestine model of CAE and towards a public spectacle; as Dominguez acknowledges, a spectacle based on *simulation*.

Dominguez grew up in Las Vegas, where he was radicalised by TV. Conscious of the nearby nuclear test site, as an adolescent he became fascinated by the Saturday monster movies. Their mix of radiation fantasies and paranoia turned Dominguez into an anti-nuclear boy scout. 'I instantly understood,' he says, 'that I was *screenal*. Simulation became more important to me than any kind of fact. I'm a deep believer in mass consensual hallucinations.'

Moving to Florida in the 1980s, he became part of the emerging CAE, although he was no longer actively involved with the group by the time CAE developed their model of electronic civil disobedience. ACT UP introduced Dominguez to protest gestures and tactics – marches, sit-ins, butt-ins (don't ask). He was involved in a successful campaign against a chain of pharmacies whose response to the AIDS crisis was to stop selling condoms. The campaign used the tactic of the 'call-in', a remorseless bombardment of phone call after phone call, saying they were loyal customers who liked shopping there – in particular to buy their condoms. They won this one.

In this shift towards a simulation model of electronic civil disobedience, Dominguez was influenced in part by the virtual mercenaries – the Panther Moderns – in William Gibson's cyberpunk novel *Neuromancer*. The Panther Moderns are a band of informational guerillas who practise 'random acts of surreal violence', crashing security systems with 'carefully prepared doses of misinformation'.[20] They exploit the gap between media representations of 'terrorism' and its actual motivations – not unlike the ways the Zapatistas were to exploit the similar gap between the stock imagery of 'guerillas' and their own actual practice, rooted in the sociopolitical conditions in Chiapas that led to their formation.

On 1 January 1994, Dominguez, now living in New York and no longer a member of CAE, saw a tantalising blip about the Zapatista uprising on CNN. When he received the EZLN's first communique that week, the Panther Moderns began to seem possible. Later that year CAE first presented their concept of electronic civil disobedience at the *Terminal Futures* conference at London's Institute of Contemporary Arts, putting the idea out in the open. After learning of the Acteal massacre in December 1997, Dominguez began trying to put electronic civil disobedience into action, sending out messages online through his contacts in the digital art world. An Italian group called the Anonymous Digital Coalition proposed a *manual* virtual sit-in. In this version, participants would just visit the target website and click 'reload/refresh' for themselves again and again; possibly the first such sit-in had been staged on 12 December 1995, in opposition to the French government's nuclear policies.[21]

A digital artist, Brett Stalbaum, heard about this and approached Dominguez with an offer to program an application, a Java applet, which would automate the sit-in, greatly increasing its effectiveness. Another acquaintance, Carmin Karasic, offered to create an online monument to the Acteal victims. The EDT was coming together. Stefan Wray had arrived in New York in 1997 to pursue a doctorate at New York University, and became the group's theorist; Wray was instrumental in developing a new model – away from the clandestine original and towards a more public, transparent simulation. Rejecting the CAE vision of electronic civil disobedience – while appropriating the term— the EDT were open about their identities, their objectives, and the tactic itself. They began to devise a Version 1.0 model of electronic civil disobedience. And this openness, as the next section shows, combined with Wray and Dominguez's accessibility and willingness to deal with the media, enabled them to create a true electronic disturbance.

The empire strikes back

On 9 September 1998 the EDT staged a FloodNet performance at the Ars Electronica festival in Linz, Austria. It was to be their longest action, their biggest action – a 24-hour virtual sit-in against three sites: President Zedillo of Mexico's home page; the Frankfurt Stock Exchange (chosen because companies listed there wanted to buy uranium mining rights in Chiapas); and the Pentagon's website. The EDT claim close to 10,000 people took part, with the FloodNet software sending around 600,000 hits per minute to each of the target sites. But this action was to draw critical, antagonistic and even hostile responses from four different fronts: anonymous agents, hacker groups, other activists and the Pentagon itself. 'At 7.30 that morning,' says Dominguez, 'I received a call from an individual with a very clear Mexican Spanish accent, who I can only say must have been a Mexican agent. They said "Is this Ricardo Dominguez?" "Yes." "Of the Electronic Disturbance Theater?" "Yes." They said in very clear Spanish, "We know where you're at, we know what you're going to do, we know where your family lives. Do not go downstairs, do not start this performance, you know that this is not a game." And they hung up. Of course, being a drama queen, I told everybody! Police were running around, people were protecting the computers. You know, part of this performance is to create this drama, what I call creating a situation where the politics of the space play out without me having to go "Oh, look at the mean bad people." People actually *see* the mean bad people doing their mean bad thing.'

For the EDT, this was only the first opposition they were to draw at the Ars Electronica event. They also angered a group of hackers, who told them that the virtual sit-in was bandwidth abuse and promised to retaliate if the sit-in continued. 'It was my first encounter,' Dominguez notes, 'with what I call the *digitally correct* community – those who hold that bandwidth is above human rights.' The virtual sit-in conflicted with the hacker ethic. If it generated enough traffic, it would block access elsewhere for people who may not have a clue what was causing their problem.

'Our consideration,' says Dominguez, 'was that in a *real* sit-in, if you have thousands of people on the Brooklyn Bridge protesting, miles and miles away you're going to have Joe Businessman going "What the fuck is going on? I can't get my coffee, I can't get to work." There is always a blockage, a backlog, where they may not find out what happened till they read the newspaper the next day. The hackers were saying, "We could take these people down in two seconds. Quick, efficient." But they would do it in a traditional hacker ethic manner – anonymity, encryption, a fairly elite kind of code. And this would represent the ability of one or two individuals, rather than the mass organic effect that we were trying to perform.'

Then they were confronted by Geert Lovink and some of the nettime critique community, who criticised FloodNet as bad activism. In their analysis, the EDT's action would disturb the capacity of a real activist network to develop. Dominguez and Wray were told to look at McSpotlight as a model of good use of the networks. 'It was a different angle than the hackers,' says Dominguez. 'But in a certain sense it did play with a set of ethics about what activism online was.' This is an ongoing debate – the alternative media model of McSpotlight, which focuses on consolidation, outreach and connection, and on its library function, versus the hit-and-run, event-driven tactical model. The EDT's response was that the virtual sit-in was just one tactic within a very long-term strategy, within a larger Zapatista call for activism. It was *supposed* to be very different from the longer-term strategic approach of a project like McSpotlight; it was an instance of *tactical* media.

When Dominguez and Wray returned to the sit-in, some five or six hours after it began, they noticed something strange. 'At the bottom of the browser window were these Java applets,' Dominguez says. 'Little coffee cups, about twenty-five of them in a row, going "*ack ack ack*". And, of course, the Netscape browser would crash, would shut down.' Email began to arrive from other participants, noting the same phenomenon. At first, the EDT assumed it was the hackers, making good on their threat to shut down FloodNet. But then they received an email from a journalist at *Wired*. 'They said, "Gentlemen, I suppose you've noticed that the sit-in is not working any more",' recalls Dominguez. 'And they said, "Do you know why?" We said, "It's the hackers." They said, "No. It's the Pentagon".'

Brett Stalbaum, the writer of the FloodNet code, identified the problem as a deliberate countermeasure – a Java applet similar to his own, called 'Hostile Applet'. A Defense Department spokesperson told *Wired* that 'Our support personnel were aware of this planned electronic civil disobedience attack and were able to take appropriate countermeasures.'[22] The spokesperson didn't go into specifics, but there is good reason to accept the EDT version of events, Hostile Applet and all. 'The DoD [Department of Defense] neither confirmed nor denied it,' points out Dorothy Denning, author of many publications on computer security issues and a policy adviser to the Clinton Administration. 'I think if they hadn't done it, they would have denied it.'

'So there you have the situation of power coming out,' says Dominguez. 'There you have *bam, bam, bam, bam*. Government intervention in terms of Mexican agents. Internal indigenous hacker response. Net critique/Net activist response. Massive-war-machine response. And in a certain sense you can see them all interlocking with each other for different reasons. We started getting calls from Harvard School of Electronic Law, and we talked with lawyers about the possibility of suing the government for violating the 1878 Posse Comitatus law that bans the use of the military in domestic law enforcement. That's why you have the FBI, the police, all these other things, because you can't use military hardware at that level. In the same way that the Pentagon is not allowed to use B-52s against New York City, they may also not be allowed to use offensive information war tactics against civilians.'

For some commentators on information warfare, such as information security consultant and author Winn Schwartau, this response to FloodNet raises serious concerns: '[T]he Pentagon's counter-attack,' writes Schwartau, 'raises questions regarding whether the U.S. Government and the military should be launching cyber-attacks within the U.S., even as a defense measure.'[23] 'US military use of cyberspace-based attacks against American citizens, were they to occur, would be a profoundly disturbing phenomenon,' says military academic and RAND Corporation analyst John Arquilla. 'However, one can "neutralise" the effects of an action without necessarily having to "retaliate".' Dorothy Denning argues that the Hostile Applet could be seen as a reasonable response. 'It is obviously controversial,' she says, 'and I think the DoD decided not to do that in the future. But the DoD is certainly entitled to defend its own computers from attack.'

The possibility of an EDT lawsuit over the Pentagon countermeasure, with its attendant potential for a media bonanza of McLibel proportions, and a new level of publicity for the Zapatista cause, is an intriguing one, but *Hacker Crackdown* author Bruce Sterling notes that the US government would probably do everything possible to avoid this. 'If you're in the government,' Sterling says, 'you don't want to give these guys publicity; they're obviously aching for a big public show trial, and if they ever get a national megaphone, that's when the situation moves from "irritant" to "debacle".'

But does it work?

Stefan Wray readily acknowledges that targeted sites have rarely, if ever, been rendered completely inaccessible during FloodNet actions. And clearly there are technical questions about whether or not a site can actually be blockaded through FloodNet – the practice of caching pages on Internet service providers' servers, for one, suggests that the targeted site would still be accessible during the virtual sit-in – and the Electronic Disturbance Theater's actions have been widely criticised as ineffective. Real hackers dismiss them as naive dabbling, boasting that they could take down servers much more efficiently. Military analysts deride hacktivists as simply a nuisance. Bruce Sterling sees hacktivism as 'basically hobbyist activity'.

But to measure the EDT's effectiveness in this way would be a mistake. FloodNet is primarily a *media* event, designed to raise awareness of the situation in Chiapas. The novelty of the tactic, combined with the mainstream media's still unsated appetite for 'Internet firsts' has enabled the EDT to garner significant amounts of coverage – major newspapers in the US, magazines (including *Wired* and *Scientific American*), ABC radio in Australia, and news outlets in Germany, Switzerland and Italy have all publicised the group's actions.[24]

'The hype generated by FloodNet has done more "damage" than the actual use of FloodNet,' says Stefan Wray. 'It was more of a coup for us to appear on the front page of the *New York Times* with a headline suggesting that Netwar had been declared against Mexico, than the actual success of FloodNet.'

If the point of the tactic is to gain media attention, then FloodNet 'works'. But it might have worked at the expense of the group's underlying objective. Much of the media coverage of the group's actions focuses on the process rather than the cause, on the technology rather than the Zapatistas. Wray acknowledges that the Zapatistas and Chiapas remain on the margins of US media, as does Mexico in general.[25] But he argues that the EDT's actions can still be seen as effective.

'I imagine in differing ways and to varying degrees our work has impacted a number of people,' Wray says. 'Remember too that we are four people, so rather than asking about effectiveness, the question, I think, needs to be asked about effectiveness relative to the number of people engaged and the amount of resources spent. Well, there are four of us. And we have basically not spent any money. Just our time. Many large companies spend huge amounts of money on advertising and marketing campaigns with the dream of getting decent media coverage. We received media coverage, and messages about the Zapatistas received media coverage, from a considerably less expensive media campaign.'[26]

Projects like FloodNet or gwbush.com can easily generate the mainstream media coverage they need, but there is a strong possibility that this is a novelty which could soon wear off. The Net itself is still novel enough for any old angle to be worth a story. The technology hasn't yet become part of the furniture, and so scarcely a day goes by without the newspapers reporting on something which has happened for-the-first-time-in-cyberspace. More often than not, such stories have little or nothing to do with the Net itself, and a lot more to do with the offline world. We might think, for instance, of the good citizens of Halfway, Oregon, who changed the name of their town to Half.com in a deal with an online retailer. 'By changing our name to Half.com,' declared the Mayor, 'our community can greatly benefit from the success of the Internet.'[27]

'A good 70% of those articles about us,' argues Ricardo Dominguez, 'even though they're about *hacking*, they always say "Electronic Disturbance Theater did it in solidarity with the Zapatistas in Chiapas, Mexico". To me that is, in a certain sense, a victory. But those spaces where it says "EDT, distributed denial of service system, FloodNet", and it mentions nothing about the Zapatistas, nothing about Chiapas, nothing about why it was done, then I'd say it's a clear loss. But I'd say a good 70% of the time, even in the *New York Times*, it was pretty specific about why we were doing these actions, what the solidarity was. I mean, they didn't go deep into who the Zapatistas were and what the conditions were, but hopefully people would respond to a certain level.'

Netwar

As the novelty of this kind of hacktivism wears off, it may follow that future actions which take cyberspace as their site will be covered more in relation to their ends rather than their means. If so, the significance of FloodNet will be in its introduction of a new way of thinking about activist tactics, tactics which may become more widely adopted and more 'effective' as media attention shifts from the technology involved to the underlying objectives and causes.

But as well as media attention, the EDT have also drawn the attention of the military establishment. RAND Corporation analyst John Arquilla is also a faculty member at the US Naval Postgraduate School, specialising in Information Warfare. He sees the virtual sit-in as a harbinger of more widespread and effective hacktivism.

'FloodNet,' Arquilla says, 'is the info age equivalent of the first sticks of bombs dropped from slow-moving Zeppelins in the Great War. Actions such as these have amounted to little, so far. However, I think we need to take a longer view. The implication, of course, is that netwar will evolve, as air war did, growing greatly in effect over time.'

For the *New York Times* to declare on its front page that the Electronic Disturbance Theater had declared 'netwar' on the Mexican government is clearly a good result for the group in PR terms.[28] But it also points to a danger, in that their actions are being framed very much in military terms. The concept of 'netwar' was developed by John Arquilla and his RAND colleague David Ronfeldt, who describe it as an emerging form of conflict, specific to what we might call the information age, and argue that the Zapatistas are its first, experimental practitioners. Their definition is worth quoting in full:

> Netwar refers to information-related conflict at a grand level between nations or societies. It means trying to disrupt, damage, or modify what a target population 'knows' or thinks it knows about itself and the world around it. A netwar may focus on public or elite opinion, or both. It may involve public diplomacy measures, propaganda and psychological campaigns, political and cultural subversion, deception of or interference with local media, infiltration of computer networks and databases, and efforts to promote a dissident or opposition movements (sic) across computer networks.[29]

So far, so much like common sense. Conflict always involves propaganda, disinformation, and attempts to control how the issues are presented and framed. Arquilla and Ronfeldt also note that 'netwar' should not be solely identified with the Internet – where email is obviously crucial in relaying information, so too are faxes, phones and print, and physical meetings, demonstrations and actions. What is distinctive about 'netwar', they argue, is this: 'The information revolution is leading to the rise of network forms of organization, whereby small, previously isolated groups can communicate, link up, and conduct coordinated joint actions as never before.'[30]

The RAND analysts cite the Vietnam War as an antecedent of 'netwar' strategies, through which the hierarchical, institutional US forces could be outmanoeuvred by more flexible, decentralised opponents. The key implication of this, combined with the increased significance of electronic information systems, is that 'Institutions can be defeated by networks. It may take networks to counter networks. The future may belong to whoever masters the network form.'[31]

'I think small groups such as ours,' says Stefan Wray, 'working in tandem with others, have the capability to find holes in the fabric of total control. As the future is constantly being woven, there will always be glitches, disruptions or snags that can be intelligently manipulated. In the end we are dealing with leveraging asymmetrical power. Increased dependence on networks may tip the balance in the sense that more people are able to enter the stage as actors, whether they be state actors or non-state actors. Modern networked warfare, as opposed to post-modern networked warfare, was limited to states, in part because of the incredible investment of capital needed and the requirement for massive armies. In networked environments, normal citizens can join in the fray.'

But the question is whether these struggles are best described using terms built around the vocabulary of warfare. Arquilla and Ronfeldt have acknowledged, for instance, that these 'social netwars' may have 'some positive effects and implications for spurring democratic reforms'. Given this point, and the participation of such groups as the Red Cross and the Catholic church in the Zapatista support campaigns, there's clearly something problematic about the 'war' part of the word 'netwar'. Why not, for example, 'netpeace'? Electronic civil disobedience, wrote CAE, 'is a nonviolent activity by its very nature, since the oppositional forces never physically confront one another'.[32] Stefan Wray has frequently used the language of 'infowar' and 'netwar' in his writings, but is increasingly uncomfortable with them.

'What we are engaged in is, and should be conceived of as, "peace-making" and not "war",' says Wray. 'When I first heard of the terms "netwar" or "cyberwar" or "infowar" I was taken by them [but I] have since come to question their use. In infowar there are no dead bodies, there are no rotting corpses, there are no blown-off heads.'

Or at least not yet. It's hardly surprising that Arquilla and Ronfeldt, as military academics and RAND analysts, see FloodNet in military terms. What's of more interest, though, is the way that media organisations like the *New York Times* iterate and amplify this kind of vocabulary in the search for a snappy cyberhook. I suggested to Arquilla that there is something problematic about this militarisation of humanitarian actions, and that the connotations of 'netwar' tend to demonise non-state actors while legitimising state actors and actions.

'While I don't think nonstate actors are demonised,' Arquilla responds, 'it is clear, according to the netwar concept, that they do become "combatants" of a sort. Thus, state efforts to control their access and influence are understandable. In the Zapatista netwar, for example, the Mexican government and military quickly understood the impact the NGOs were having, and took effective steps to control their physical access to the area. They were less successful in preventing electronic access. The bottom line, though, is that the netwar concept does herald a new blurring of the line between those participating in a conflict, and those whose motives are purely humanitarian.'

This 'bottom line' is the worrying part. If, as Arquilla suggests, the line between conflict and humanitarian action is blurred, then we need to tread it warily; we need to be conscious of the politics of naming and control. Once metaphors such as 'netwar' take hold in public discussions, they have a powerful effect on the directions those discussions can take. For example, as media academic Philip Schlesinger and his colleagues have argued in relation to definitions of 'terrorism', such debates are more than word games:

> Real political outcomes are at stake... if the public can be persuaded that the state is right, this helps mobilise support for transferring resources from welfare to security. Language matters, and how the media use language matters.[33]

Schlesinger et al. identify four perspectives on terrorism commonly found in the media; prominence in media coverage generally goes to the 'official' perspective:

> the set of views, arguments, explanations and policy suggestions advanced by those who speak for the state. The key users of these official definitions of terrorism are government ministers, conservative politicians, and top security personnel. Given their high status as news sources their opinions are assured a prominent place in media coverage.[34]

As with coverage of hacking, this 'official perspective' abstracts an act from its political context by emphasising its criminality. And such politics of naming and control are also central to the tensions between the Net Versions 1.0 and 2.0, as we'll see below.

As Wray's uneasiness with 'netwar' vocabulary shows, the EDT are very aware of these issues of naming and control, of the power struggles that are acted out through language itself, the domain of soft power. In a presentation to the US National Security Agency in September 1999, Wray pointed out that the conference program had renamed his group 'the Electronic *Disruption* Theater' and described the Zapatistas as a 'sect'. While these could have been simple errors, Wray argued that they may also have represented 'an attempt to recategorize who we are into a framework that is understandable to the national security mindset'.[35] Again, this is not a trivial issue. The ways in which actions are framed and described, the motives attributed, meanings sought and implied, are a fundamental power struggle. All of which can make the word 'hacktivism' a serious problem for anyone to whom it might be applied.

'Hacktivism' is a superficially attractive term which has encouraged reporters to describe FloodNet as an example of an emergent politically motivated hacking. The assumption here is that hacking has previously been entirely apolitical, something which derives more from the nerdy stereotype of the Hollywood hacker than from the politics of information access and control which have underpinned much hacker activity. Insofar as 'information wants to be free' expresses an informational struggle, then hacking has always been essentially political on some levels.[36] This is clear from a project such as Peacefire, which offers instructions in how to disable content filter software – the kind of product frequently used by parents in attempts to control what areas of the Net their kids can get access to: 'It's not a crime to be smarter than your parents,' reads a slogan on the site.[37] But, as we've seen, more sceptical perspectives are possible: 'It was never unusual for computer cracking to have a political tinge,' says Bruce Sterling, author of *The Hacker Crackdown*. 'Even lowly phone phreaks putting washers into payphones used to employ Vietnam War resistance as an ethical excuse. There's always been a vague bohemian self-righteousness associated with computer countercultures.'

While he acknowledges the risks in the term, Ricardo Dominguez says that because of their association with 'hacktivism', the EDT were invited to hacker gatherings – H2K in New York, Def Con in Las Vegas. And they were then able to speak to hackers directly, to try to find a way of overcoming their earlier inability – and the inability of CAE before them – to politicise hacker groups. 'Our presentations weren't about software,' he says. 'They weren't about secret phreaking methods. They were "Let's look at what politics means in terms of civil disobedience". All these hackers were sitting in a room looking at an 1848 text by Thoreau. And they were taking notes! Which to me was utterly amazing. So while I feel the term "hacktivism" doesn't really signify the real conditions, it has implemented a swarming effect. And it would just be too difficult to try to counter it.'

Dominguez isn't a big fan of the word 'hacktivism'. He points out it was coined by a reporter and argues that 'electronic civil disobedience' is much more specific. But the EDT made a pragmatic decision to exploit the new term and its media interest; to try to find ways of using it to articulate what they were doing, rather than struggle against it. 'Hacktivism,' says Dominguez, 'is still a very empty term. But at the same time, empty terms do allow a certain kind of energetic invention, because people try to fill them. So, many kinds of things begin to aggregate into filling the burrito of hacktivism.'

Filling the burrito

One contribution to this burrito was the planned end of the FloodNet project. Late in 1998 an Australian Aboriginal activist group from Woomera contacted the EDT, asking if they would help them to stage an action. The EDT's initial response was that 'we only help the Zapatistas'. But as they thought about it, they turned this request into a solution to the problem of how to end the project. The EDT made the FloodNet code freely available for download at midnight on 1 January 1999 (although this was supposed to be the end, the EDT later reactivated FloodNet in the 'digital storm' of August 1999, and in what became known as the Toywar). This download, in its openness and transparency, in its enabling of new horizontal connections, was a very Version 1.0 gesture. In no time at all, Queer Nation used it to stage an attack on godhatesfags.com, and throughout 1999 other groups also used it: the Electrohippies, for example, who staged a virtual sit-in as part of Seattle N30.

But the burrito is filling up in other ways besides virtual sit-ins.[38] Some tactics that come under the hacktivism banner are similar to FloodNet's denial-of-service technique. An example is the *email bomb*, another unfortunate choice of term, which describes a kind of cross between a virtual sit-in and a petition. In these actions, emails flood in to overwhelm the system's server and to demonstrate the extent of support for the cause in question; these can be automated as a kind of chain letter, so that participants can send a message – or lots of them – just by clicking on a link. In 1998 ethnic Tamil groups carried out perhaps the first such hacktivist email action, against Sri Lankan embassies. It aimed both to draw publicity and to overwhelm the embassies' email systems, and was successful on the first count at least.[39]

In Australia, *Workers Online*, the web zine of the NSW Labor Council, set up an email action of this kind in June 2001. Subscribers were sent an invitation to visit a link and send an email to the NSW parliament in response to legislation on workers' compensation.[40] Within two hours, a reported 13,000 emails had jammed the server. In fact, there was so much traffic that the state government blocked email coming from certain domains, including the Labor Council. As *Workers Online* pointed out, this was an ironic response – the Premier had criticised a physical blockade earlier that week, yet the government response to the email action was to erect a virtual blockade of its own.

Other tactics, though, are closer still to the popular image of hacking. Like the virtual sit-in, some are updates of older tactics. The web is alive with upscale graffiti artists, for instance, who've defaced or amended the web pages of groups as diverse as Greenpeace, the Spice Girls and the Ku Klux Klan. In September 1998 dozens of Indonesian sites had 'Free East Timor' added by Portuguese hacktivists.[41] In October 2000 a website operated by Hezbollah had a Star of David added to its front page by pro-Israeli hacktivists.[42] And in October 2001 a group called GForce Pakistan left a pro-bin Laden message on the website of one US government agency.[43] Hacker site Attrition.org states that more than 5000 such graffiti actions took place in the year 2000 (although many of these would not have been driven by activist agendas so much as by kids showing that they could do it).[44]

Another hacktivist tactic is a kind of hijack. For example, as part of the protests surrounding the World Economic Forum in Melbourne on 11 September 2000, the domain name address of Nike was redirected, sending surfers who typed in 'nike' to the S11 website. Other supporters registered misspellings of the word 'Olympics', redirecting anyone who typed, for instance, 'olympisc.com' to the S11 site – making it for a time one of the most popular sites in Australia.[45] And in March 2001 hackers redirected Arab group Hamas's web page to porn sites 'teenjuice.com' and 'hotmotel.com'.[46]

Still other tactics make more use of the technical properties of the Net (if there's anywhere we might expect to see entirely *new* tactics develop it's in the hacker-activism realm). One example would be the electronic saboteurs who shut down the communications systems of the Australian Republican Movement in the lead-up to the constitutional referendum of November 1999.[47] Another would be the hackers associated with the group behind a site called Condemned.org, who were reported in January 2000 to have broken into the servers of a number of child porn sites and erased their hard drives.[48] And then there would be political uses of viruses and worms. One Israeli teenager claimed to have destroyed an Iraqi government site by sending a virus in an email attachment[49] (although in terms of drawing publicity, virus hoaxes, which are a kind of virus themselves, could be just as effective as the real thing).

So in this context, *is* FloodNet hacking in any meaningful sense? EDT member Carmin Karasic points out that 'FloodNet never accessed or destroyed any data, nor tampered with security, nor changed websites, nor crashed servers.'[50] And if it's not, what do the EDT stand to lose from their association with hacking? Hacking, after all, is scary for many people.

Yet without hackers there would *be* no Internet – at least not if we're using Steven Levy's original sense of the word 'hacker'. And if the EDT are hackers at all, then it's in Levy's sense, which he applied to the innovators and designers of the early computer industry. In Levy's description, these early hackers were very much about a Version 1.0 view of computing: 'It was a philosophy of sharing, openness, decentralization, and getting your hands on machines at any cost – to improve the machines, and to improve the world.'[51] A 'hack' was an elegant solution to a technological problem; more than that, it had to be, as Levy says, 'imbued with innovation, style, and technical virtuosity'.[52] In Levy's usage, hacking was about improving systems rather than crashing them; about sharing information rather than stealing or changing it. The early hackers made computer breakthroughs, not break-ins.

But the meaning of the word changed fast. In *The Hacker Crackdown*, Bruce Sterling details the 1990 co-ordinated arrests and show trials in the US, which led directly to the formation of the Electronic Frontier Foundation. Sterling sees the real struggle in the early hacker show trials as one over control of *language*: 'The real struggle was over the control of telco language, the control of telco knowledge. It was a struggle to defend the social "membrane of differentiation" that forms the wall of the telco community's ivory tower.'[53] A struggle, in other words, over inclusion and exclusion, and over naming and control. And a struggle which created a distinction between what communication scholar Douglas Thomas calls 'old school' and 'new school' hackers.[54] The old school were the heroes of Levy's book: the innovators at MIT in the 1950s and 1960s and in California in the 1970s, typified by Steven Wozniak, creator of the Apple. The new school are the ones who go to jail.

This new school was created on several fronts. The *criminalisation* of hacking took place in the courts.[55] But it needed the *demonisation* of hacking, which took place in the media: on the one hand, through fictional depictions of hacker activity, particularly in Hollywood movies such as *War Games*, where a high school student almost starts World War III from his suburban bedroom; *Die Hard 2*, where criminals hack an air traffic control system and crash a jet filled with passengers; and *The Net*, where Sandra Bullock's character's identity is erased online. And on the other hand, the media were also filled with lurid hyperbole surrounding real-life hacker incidents – 'Invasion of the Data Snatchers!' yelled *Time* magazine in 1988.[56] Such coverage made folk devils of the likes of Robert Morris;[57] Kevin Poulsen;[58] and, of course, Kevin Mitnick, who was compared to Darth Vader and described as 'a real electronic Hannibal Lecter'.[59]

The creation of 'new school' hackers was a classic moral panic – an expression, in the media, of cultural negotiation; of working out where the boundaries lie through processes of inclusion and exclusion.[60] A moral panic, like that over hackers, is a moment when this process bubbles up to the surface of the media.[61] The key thing to a moral panic is that it presents 'us' as being under some kind of threat from 'them'. But who 'they' are tends to be different each time – and who 'we' are can be problematic as well. We could try to offer a positive definition of who we are, but we can also do this negatively – we can define ourselves by reference to others. 'We' are different from 'them', and this becomes part of the explanation of who 'we' are. And by excluding those who're not in our big cultural gang, we define ourselves negatively – by exclusion.

In his book *The Politics of Pictures*, cultural studies academic John Hartley explores these processes of inclusion and exclusion. He draws on Benedict Anderson's idea that the nation is an 'imagined community' – that in a complex modern society the only contact we have with most of the other people in it is *symbolic*: we know each other through our media. And the media work to generate and sustain that sense of collective identity. They do this through stories which represent people or events or values as being 'ours' or 'theirs'. Readers, writes Hartley, are 'encouraged by each newspaper . . . to see the news as part of their own identity, while the news strives to identify with them'.[62] Those excluded tend to get treated in routine ways – they'll be stereotyped and simplified so that they are all seen to be like each other (and hence different from 'us'). They will, for instance, be painted as 'data snatchers' or 'electronic Hannibal Lecters'.

But if the creation of the new school hackers was a moral panic, then who was behind it?[63] We might begin by asking who benefits from demonising new school hackers. Governments, with a vested interest in developing information infrastructure? Security agencies, with a vested interest in controlling what they see as new crimes?[64] Or the old school hackers themselves, who, as Thomas puts it, 'went corporate':

> The first thing Jobs and Wozniak did, for example, was make the Apple proprietary (which may well lead to Apple's ultimate demise).

> They took the hacker ethic, 'information wants to be free' and turned it on its head. Others followed suit. The only person not to sell out was Bill Gates – he was corporate from the very beginning.[65]

This explanation carries a certain weight; cultural critic Andrew Ross, for example, has pointed out that hacker and virus scares 'resulted in a windfall for software producers'.[66] But perhaps, in asking who benefits, the best answer is *all of the above*. By 1990, as computing became more deeply implicated in the whole fabric of society and the economy, hacking had become, as Bruce Sterling notes, 'too important to be left to the hackers'.[67]

The creation of new school hackers can be seen as another aspect of the Version 2.0 struggle for commercial dominance of cyberspace. The original sense of 'hacker' and the open-system hacker ethic just didn't fit with the consolidation of information technology at the centre of the 'new economy'. And the marginalisation and demonisation of the hacker ethic, in which information famously 'wants to be free', was a necessary step in consolidating a worldview in which information would prefer to be paid for.

All of which means that promoting an emergent cyberspatial politics as 'hacktivism' means dealing with the baggage of the 'hack' part of the word. This may make it all too easy for hacktivism, or electronic civil disobedience, to be shot down, as Stefan Wray acknowledges.

'The state will have less of a monopoly on warfare than has been the case historically,' Wray says. 'This, in part, I think, is why they are so keen on demonising hacktivism as a form of terrorism. So long as state action can be sanctified and non-state action can be vilified, then they will stay on top.'

One challenge for activists, then, is not just to formulate new strategies and tactics appropriate to a shifting mediascape, but to recognise the ongoing need to create a vocabulary for discussing those strategies which is not fatally compromised from the start.

Toywar

29 November 1999: Online toy shop ETOYS *is granted an injunction which bans European digital art collective* ETOY *from using their own website. The addresses are certainly similar – www.etoy.com for art; www.etoys.com for, well, toys. But etoy is not a parody site in the vein of gwbush.com. The art group had registered, and begun using, their address in 1995, nearly two years before the toy company was even founded.*[68] *Worried, apparently, that its customers would have trouble typing the letter 's', the retailer first proposes to buy the artists' web address, offering a reported US$516,000 in cash and stock options. When the artists decline, eToys cranks out its law firm, crying trademark infringement, arguing that the situation will cause confusion for shoppers, and throwing in the usual out-of-nowhere complaints about violence and porn.*

They couldn't have picked a worse target: etoy – slogan 'leaving reality behind' – specialise in creating 'insane digital tools', 'revolutionary incubations', 'digital hijacks' and other such fun stuff. The group is well established in digital art circles, and has close links with all kinds of experimental art networks, media activists and culture jammers. It would have been hard to find anyone better qualified to make a media event out of turning the tables on the toy shop and exploiting their suit. Support sites were set up, counter-suits filed, investor groups were lobbied online; high-profile Net commentators John Perry Barlow and Douglas Rushkoff criticised the toy shop in the press. The EDT, along with ®™ark, Fakeshop and etoy themselves, mounted a '12 Days of Christmas' action, with the EDT deploying FloodNet against the eToys website.

'You have to remember, 1999 was the year of the dot com boom,' Dominguez says, 'where billions and billions were being produced in this artificial bubble. eToys.com was one of these mega-firms that existed only online. [The Internet] was considered an international network where nobody really has control. [But] using their wealth, using American law, they showed that [that]... was quite wrong. This seemed to us an outrage, not only against domain politics [and] network politics, but they were attacking Net art. But what they did not realise was that we have a long tactical media tradition – we were not silent.'

During the activist swarm against eToys, its share price – which had been rising until the ugly publicity began – plunged from close to US$70 to less than $20. Another element of this campaign was what Dominguez calls 'psychological operations' – activists would go into online trading boards for eToys stocks and agitate investors. 'Obviously [what happened] could have been an early weathervane thing of what was coming a few months later,' he says, referring to the NASDAQ 'correction'. 'But it happened dramatically... their stocks were going down at exactly the same time that we were doing the action. And we were hitting the trading websites, going "Look at this! Why? Because they're stupid, because they're spending your money on some stupid etoy Net art thing." So all of this aggregated to a very powerful gesture. And on 15 January they relented, they said we'll pay you back, we'll never bother another Net art community again. And of course now they're dead. They're completely dead.'[69]

Staging an action against the Pentagon is one thing. But messing with the share price of a toy shop? That's serious. Among other things, the Toywar produced a whole new round of hacker scare stories. CNN jumped out from behind the couch to yell boo about 'ping-launching zombie machines', as part of a special section of their website headed 'Insurgency on the Internet'.[70] ('Insurgency?' asked ®™ark. 'Where's the regime?') For the EDT, the electronic disturbance this time was much more straightforward than in the Pentagon event. On the second day of the sit-in, Dominguez woke up to find that the EDT site had disappeared.

'On the first day of the action,' says Dominguez, 'eToys said "No, nothing's wrong, nothing's wrong with our servers, we're perfect, we're up." On the second day I woke up and there was no Thing.net, there was no EDT, there was no ®™ark, there was nobody. What we discovered was that eToys.com had contacted the FBI, who then contacted our router – the bandwidth barons. If in the frontiers you had railroad barons, now we have bandwidth barons. And they shut the Thing, which is an Internet service provider.'

Later that day, the Thing were told by their infrastructure routers that their services would be restored if Dominguez's EDT site was removed. Dominguez complied, not wanting to hold anyone else hostage. 'By that night,' he says, 'I was on CNN, I was again in the *New York Times.* That's one of the things about these situations – *you want to create a situation where power, no matter what it does, fails.* If they shut you down, they fail. If they shoot offensive information war weapons, they fail. I think that's the most positive thing for activists – that no matter what response power does, they fail. And it only ups the information distribution of what you're doing.'

Faced with a situation where Version 2.0 power had shown itself, and shown the extent of the commercialisation of cyberspace, Dominguez did the only thing that he thought could be done. He put the contents of the EDT site up for auction. Except nobody bought it, so around three months later he quietly put it back up, where it remains. The Toywar, Dominguez argues, shows both the limits and the possibilities of the EDT's approach.

'The Toywar,' says Dominguez, 'showed that electronic civil disobedience has a certain symbolic efficacy against power. With the Mexican government, no matter what you do to their website, you're not going to disturb their tanks, their missiles. No matter how much you disturb Nike's website, you're not going to disturb their stores, because they have real exchange power on the ground. But with eToys, they were completely virtual, they had no real representation. So you saw that electronic civil disobedience is an extremely useful tool when dealing with a *virtual* organisation, [an organisation] that is *only* virtual. And perhaps as we become more virtual in the future, electronic civil disobedience may become a more useful tool. Not just as a symbolic efficacy, but as a means for really leveraging activist concerns.'

Coming up?

In their latest book, CAE come down against the transparency of the EDT, arguing that electronic civil disobedience is 'an underground activity that should be kept out of the public/popular sphere (as in the hacker tradition) and the eye of the media'.[71] CAE argue that the kind of media manipulation attempted by the EDT is another dead tactic, because all potential opponents are now ready and willing to engage on the media terrain – writing the occasional question mark into the body of media discourse won't be enough. So, in the spirit of tactical media, CAE call for an end to electronic civil disobedience:

> It has already been sold for fifteen minutes of fame, and is fueling a new round of cyberhype, but e-activists can bring a halt to this current media event by supplying nothing more.[72]

In contrast to this view, Ricardo Dominguez argues that new coalitions are coming together. With the participation of the *2600* hacker group in street actions at the Republican Party conference in Philadelphia, he no longer thinks that the hacker community will necessarily remain apolitical. If hacking was political in a limited, virtual sense before, Dominguez argues that it's becoming more genuinely engaged now, and that this would not have been possible with a clandestine model of electronic civil disobedience.

'I think it's very important for us, in that we were a transparent performance, that we do get dispersed media. And the level of dispersed media wasn't just the *New York Times* – it was punk magazines, it was hip-hop magazines, it was high-end German art magazines. To us that was very important, because if you stay within just the hyper-academic theoretical condition, it's not going to get very far. If you stay within just the deep core political, it's not going to get very far. But our performance was about *getting out there*. Because the stage was the media. What did occur out of it was we saw the movements of the streets linking with the networks. Because we did a transparent gesture, we were invited by government agencies from the left and the right to come speak to them, because they were traumatised too – they're going, "What the hell is this? It's not cyber-crime, it's not cyber-terrorism." We tried to convince them that it was something completely other than their expectations. And if we *had* done it secretly, I don't think it would have [had] the same response.'

While a bunch of computer fans on a street in Philadelphia might not seem like a big deal, it's a suggestive Version 1.0 instance. Hackers, the indigenous culture of cyberspace, have been invited to build a bridge over from their purist digital realm. Building such bridges, making new connections, forging new coalitions, is what most of the projects in this book have in common – from LabourStart's international union links and Huaren's involvement in global mobilisations, through the Help B92 campaign and the Indymedia movement, to ®™ark's mutual funds. The more bridges the better, if anyone is to take the possibilities of cyberspace and move them into the world that most of us still live in.

Where Do You Want To Go Tomorrow?

When the *Titanic* hit its iceberg, Version 1.0 of radio went down with the ship. As sociologist Eszter Hargittai argues, radio turned out to be implicated in the disaster in ways that poisoned popular opinion against an unregulated, open system. A nearby ship could have saved passengers, for example, but its radio wasn't switched on. Worse, the babel of messages from amateurs produced conflicting news about whether or not the ship was safe. The resulting mood in the public and the press made it easy for the US government to regulate the airwaves. Once in place, this regulation enabled the development of commercial broadcast networks, which changed radio from point-to-point to broadcast technology.[1]

When American Airlines Flight 11 and United Flight 175 hit the World Trade Center on 11 September 2001, they may have been, among many other things, the iceberg of the Internet. As I write this six months later, it's still too early to see the long-term consequences for the Internet of the 'War on Terrorism'. But it doesn't look good. Here it all comes – real-time government interception of email. Police access to web surfing records without warrants. An end to privacy in favour of 'security'. With liberty and wiretaps for all. Version 1.0 may be out of step with the times.[2]

But hang on, didn't we hear that the Net was indestructible? Didn't we hear it was beyond censorship, that it treated such interference as damage, and would 'route around' attempts to control or constrain it? Didn't we hear that the Net was somehow stronger and smarter than us? We sure did. But it's not true any more, if it ever was. Perhaps it *was* all just technological determinism. Because as legal academic Lawrence Lessig shows in his book *Code*, the Net can be censored, constrained and controlled only too easily. Not through the gesture politics of 'adult content' restrictions, but through remaking its infrastructure. Through rewriting the code that makes it function. By adding to, or subtracting from, the infrastructure underpinning and overlaying the Internet. Through changing the *inbuilt politics* of the Net.

MP3 files can be encoded to prevent their distribution. Software can be programmed to stop working if it's not registered to you. Websites can be configured to deny access to those who are not paying subscribers, or who don't have their browser cookies enabled, or whose machine otherwise identifies them as undesirable. None of these is sci-fi; all are with us already. And they could be just the start. Lessig describes 'a future of control in large part exercised by technologies of commerce, backed by the rule of law'.[3] Business will continue to push for more Version 2.0 innovations, to make e-commerce easier to do and harder to avoid. Governments will back this, and will also make use of the increased certification and credentialling necessary to function online. On the Internet, said one dog to another in the legendary cartoon, nobody knows you're a dog. Not any more – on this Version 2.0 Net, not only will everybody know it's a dog, they'll have survey data on its favourite kind of dog food (and it won't get it without a credit card).

So – it was fun while it lasted, but perhaps it's time to log off and see what's on TV? Not so fast. Lessig concludes that all of us need to become active citizens of cyberspace – not just customers. If we decide we want a Version 2.0 closed system, with extra TV shows all round, then fine. So long as we decide this and don't have it decided for us. Attempting prophecy about technology is a loser's game – there's no future in it – but chances are that activism *about* the Internet is going to increase. Campaigns in support of the Net as an open system; against commercialisation, against censorship.[4]

If the people interviewed in these pages have anything in common, it's that they're all very much active citizens of cyberspace. I said in the introduction that I never intended this book to be the last word on any of this; it's only a contribution to an ongoing public conversation. To that end, I've tried to let the people involved in these campaigns speak for themselves as much as possible; to explain their own motivations and ideas. And I also want to give them the last word. While researching this book, I've asked those involved what advice they would give to others who want to use the Net in trying to effect some kind of social, cultural or political change. I'll end the book with some of these responses. It is, I hope, a Version 1.0 ending.

Some, like Esther Dyson, counsel optimism. 'Individuals and small businesses get access to a lot of information which empowers them,' she says. 'The examples are in the lives of the millions of individuals using the Net, right now, to get what they want, to express themselves, to communicate with one another without a central intermediary. But because this power goes to millions of people individually (and is over their own lives and not over others' lives), each instance doesn't seem dramatic. It just seems like lots of anecdotes – but they add up to a significant shift in power.'

'The web has created a space for people to experiment with new styles not motivated by sales,' says Matthew Arnison. 'When Indymedia has an audience of people energised by information, they don't wait for the experts to do something. They don't even wait for the experts to tell them what to do. They get in there and suggest things and then go out and do it themselves.'

Others, like McLibel defendant Dave Morris, are more cautious. 'The Internet,' Morris says, 'like most supposedly "labour-saving" devices, takes up a lot of people's time and energy. It is a mixed blessing, like all technology. How much resources – materials, labour (often cheap labour in poor countries) and mental effort – are being sucked into the production of such technology? With half the world having never used a telephone, modern technological communications, especially the Internet, can easily be elitist, and at the same time provide an illusion of interaction between ordinary people. In the same way that cars seem to promote movement but actually inhibit it, technological communications are tending to inhibit, sideline and replace face-to-face contacts and association, whether one-to-one or in communal spaces.'

But most offer a combination of optimistic encouragement and sober caveats. 'I have the feeling that we are running out of time,' says Geert Lovink. 'Soon the Net will be a closed mass medium with little or no room for new players. But we can then begin to build parallel networks, underground systems, somewhere in the margins. Wonderful subcultures will blossom, so please do not become depressed. There is still enough time to create parallel, independent infrastructures in which cyberculture can reinvent itself.'

I've grouped the following selection of responses under headings which reflect recurring themes. Several say activists should think tactically; others advise a strategic take, particularly with regard to networking. Some point out that it's crucial to be clear about what you want to achieve. Most urge activists not to put too much faith in the Net; others encourage activists to have faith in themselves.

Act tactically

Some advise a tactical approach – seizing unguarded moments. 'The Internet,' says Veran Matic of B92, 'is not a kind of Forbidden City, a ghetto for the sacred, but should be regarded as a kiosk full of various offers. It could be said that the Internet is the perfect guerilla weapon. "Hit and run" constitutes its main role.' 'Look for niches,' says Stefan Wray of the Electronic Disturbance Theater, 'opportunities to exploit in ways not thought of before.'

Others offer advice from their own very specific expertise. Frank Guerrero of ®™ark, for instance, makes some points about cultural sabotage. 'Don't let the legal thugs hired by corporations bully you around,' he says. 'Usually their letters are empty threats designed for intimidation. And if you are concerned about top tip #1, then register your new domain under an assumed name – give a PO box address rather than your own, and set up voicemail for phone. And learn to dance.'

Think strategically

Complementing the tactical take on things, some stress the importance of long-term strategic goals, though obviously these vary. Here are three.

'I am on a crusade,' says Geert Lovink, 'to increase what I call "economic competence". Any kind of business is a good one, because it will give you a good inside view on how big websites are being operated and what it means to get millions of clicks on your site each day. Economics is boring, I know. And I should reject it, being an anarchy-pragmatist. But I don't. If we still have the naive idea that an open and diverse cyberculture can somewhat influence the course technology is taking, we now have to start up businesses and pollute the concepts used under the umbrella of the term "New Economy".'

'In the transitional societies,' says Veran Matic, 'the Internet can really help to overcome the gap between developing and developed countries; to articulate and accept basic information of world progress more quickly. The Internet entirely changes one's life philosophy and culture. In a society full of illiterate people, where the cultural model is pretty closed, this is very important.'

'I think we have to learn from the ground,' says Ricardo Dominguez, of the Electronic Disturbance Theater. 'We have to go back to more traditional spaces. Especially off-grid spaces, people who're not connected, like the Zapatistas. They have no electricity, they have no computers. Because those populations, which are invisible, which are silent, are very easily left out from the connectivity of power, the grid. If we as activists and hackers only consider that which is on the grid as the valuable source, then I think we really lose what the Zapatistas have taught us about developing long-term political strategies. What we did is only a tactic – a simple tactic, a tactic that is effective on some occasions. But it has to be within a larger strategy. And that can only come from depth of knowledge, depth of understanding, and, most importantly, out of what the Zapatistas call the *encounter* and the *dialogue*. And you have to remember that from that encounter and that dialogue, what you want to accomplish may not come out. But something else might come out.'

Be clear about what you want

'Be strategic and pragmatic,' advises Zack Exley, of gwbush.com. 'In other words, work to get something done, not just to "do something".' 'Ground it in a real problem or issue in a community you have a stake in offline,' says Howard Rheingold. 'Use online media to organise, discuss, and disseminate information, within the framework of a well-considered political strategy.' Rheingold's final point may be the single most important one in the book – 'Don't forget to get together in the physical world'.

'It's very important,' says Ricardo Dominguez, 'that one be very specific about what is the issue that you're working on. What is the singularity of the locale? Because the answers to how to use the thing have to come not from the networks outside, but from the specificity of what you're doing. The most important lesson to be learned from hacktivism is that it's a long-term thing – one hack does not an activist make, nor does it shift a government or anything. Sometimes it takes ten years, twenty years, of very long-term depth of analysis. The action that you want to do may not fit, may not really help what it is you're trying to help. [The thing that does work] might be something that you hadn't thought of, because you were hot to do *this*. All of a sudden you might analyse it, talk to the people that you're trying to have a dialogue with, and they might go, well, no. But out of the dialogue may come another action. So I think it's very important for young activists, hacktivists, to always make sure that they understand the specificity of what they're trying to articulate. And that they become aware not to have the media, or the medium, define the articulation. You have to have the local, the singularity, define the way the medium will play itself out. Or whether the medium will be used.'

Don't have too much faith in the Net . . .

This is the single most common recurring theme. Almost everyone makes some version of this point, though with important variations. For some, 'don't rely on the Net' means being aware that many people aren't online. For others, it means remembering the importance of having an integrated media strategy.

'The trouble with the Net,' says Mongrel's Matthew Fuller, 'is that people have to have a greater range of bullshit detectors than they have to use to deal with mainstream media. Learning how to put these [detectors] together from the massive amount of customisable pieces available – and also having to learn to train them on themselves – takes enough time. Once the denormaliser starts running though, there's no stopping it.'

'Be aware of the Net's limitations,' says Gabrielle Kuiper, of Sydney Indymedia and Active Sydney. 'Don't use it as your primary form of communication. If you are spending your whole day staring at a computer screen and not getting out there and having face-to-face conversations, then it is unlikely you are changing much. It is a valuable communications technology, but it is no substitute for in-the-flesh human communication in all its messy three-dimensional sight, sound, touch, taste and smell interaction.'

'Don't expect the Net to do more than it can,' says Eric Lee, of LabourStart. 'No authoritarian regime has yet been toppled by a Java applet. It takes men and women engaged in real struggle – including strikes, boycotts, demonstrations, and so on – to change the world.'

'Be truly multimedia,' says Kalle Lasn, of Adbusters. 'I think if you want to launch successful social marketing campaigns, then you have to have radio, you have to have graphic materials like print ads or spoof ads, subvertisements or posters. You need all that paraphernalia, right down to bumper stickers and whatnot. And then, if you can afford it, you should get yourself some kind of TV spot that takes the very essence of your message and condenses it into 30 or 60 seconds. And then you have to start provoking the TV stations in your area to air it, and play that whole multimedia game as best you can. So that's my advice, not to expect the Internet by itself to pull off miracles, because it won't. We believe the activists of the future will be new media players. They will play the whole spectrum.'

. . . but do have faith in yourselves

'Ground yourself in hope,' says Mitch Kapor, of the Electronic Frontier Foundation. 'Keep your eyes wide open as to how things actually are and stick close to your own expertise.' 'Know and respect the people you're organising or communicating to,' says Zack Exley. 'Make sure you are motivated primarily by a desire to make life better for the billions who are in agony, and not by ego.'

'If you have the arguments,' says Geert Lovink, 'if you've done the research, and if you've worked with people – if you are networked in communities and groups – that's the main thing, and the technology is secondary.'

'The Internet has got humans ringing like a bell,' says Matthew Arnison. 'Just like the success of the telephone, people want to be creative and communicate and tell stories as well as listen. Just because TV is successful doesn't mean that all we want to ever do is watch. And the evidence is that people are turning off the TV to spend time on the Net. Of course many corporations are trying to bend the Internet into a tool for consumption rather than creativity. But you can only bend human nature and technology so far. Very often these attempts fail miserably.'

'Use your brain and your imagination,' says Jessy, from McSpotlight, 'to come up with new ways of doing things faster, doing things funnier. Wit is one of the most important things. McSpotlight is not made with money – it's made with imagination, and that's why people go there and are interested in it. So the Internet is a level playing field in that sense. It's not the boys with the most money who make the difference – it's the boys and girls with the most imagination.'

Notes

Introduction

1 Bertolt Brecht (1983) [1930] 'Radio as a Means of Communication', in Armand Mattelart and Seth Siegelaub (eds) *Communication and Class Struggle, volume 2*, International General, New York, pp. 169–71. Brecht's observation was echoed 40 years later by Hans Magnus Enzensberger with regard to television and film. See his 'Constituents of a Theory of the Media', anthologised in, inter alia, John Thornton Caldwell (ed.) (2000) *Theories of the New Media: a Historical Perspective*, Athlone Press, London, pp. 51–76.

2 William S. Dutton (1929) 'Minute Men of the Air', originally read into the US *Congressional Record*, October 21, vol. 71, pt. 5, p. 4699. Available online in Stephen Duncombe and Andrew Mattson (eds) (1996) *Primary Documents*, no. 5, http://home.earthlink.Net/~srduncombe/radio.htm.

3 Steve Lohr (1996) 'The Great Unplugged Masses Confront the Future', *New York Times*, 21 April, section 4, pp. 1, 6. Lohr is quoting an unspecified issue of *Radio Broadcast* magazine from 1922.

4 Glenn Frank (1935) 'Radio as an Educational Force', *Annals of the American Academy*, January, pp. 119–22. Available online in Duncombe and Mattson, *Primary Documents*.

5 See, for example, Robert W. McChesney (1996) 'The Internet and U.S. Communication Policy-Making in Historical and Critical Perspective', *Journal of Communication*, vol. 46, no. 1, pp. 98–125.

6 Although there are also important differences. The Internet was directly funded by the US government from the beginning, while radio was developed by commercial entrepreneurs who made early licensing agreements with the military in several countries: Marconi's wireless telegraphy technology, for instance, was licensed to the UK navy in 1903, five years after its first military use in the Boer War. Unlike the early Internet, this was a commercial venture which was aided by military investment. In the case of radio, it took the *Titanic* disaster and then World War I for the US government to conceive of a need for regulation. Once in place, this regulation enabled the development of commercial broadcast networks, which changed radio from point-to-point to broadcast technology. See Patrice Flichy (1995) *Dynamics of Modern Communication: the Shaping and Impact of New Communication Technologies*,

Sage, London; and Eszter Hargittai (2001) 'Radio's Lessons for the Internet', *Communications of the ACM*, vol. 43, no. 1, January, pp. 51–57.

7 Jean-Francois Lyotard (1984) *The Postmodern Condition: a Report on Knowledge*, Manchester University Press, Manchester.

8 Quoted in 'E-lectioneering' (1995) *The Economist*, 17 June, p. 21.

9 John Markoff (1995) 'If Medium Is the Message, the Message Is the Web', *New York Times*, 20 November, pp. A1, D5.

10 Nicholas Negroponte (1995) *Being Digital*, Hodder & Stoughton, Rydalmere, NSW, p. 7.

11 Howard Rheingold (1994) *The Virtual Community*, Minerva, London, p. 14. The full text of this book is online at http://www.rheingold.com/vc/book.

12 William Gibson (1984) *Neuromancer*, HarperCollins, London, p. 67.

13 Mitch Kapor (1993) 'Where is the Digital Highway Really Heading?', *Wired* 1.03, March, http://www.wired.com/wired/archive/1.03/kapor.on.nii.html.

14 John Perry Barlow (1994) 'Jackboots on the Infobahn', *Wired* 2.04, April, http://www.wired.com/wired/archive/2.04/privacy.barlow.html. The Electronic Frontier Foundation, of which Kapor and Barlow are co-founders, is at http://www.eff.org.

15 Chip Bayers (2000) 'Capitalist Econstruction', *Wired* 8.03, March, http://www.wired.com/wired/archive/8.03/markets.html.

16 David Resnick (1997) 'Politics on the Internet: the Normalization of Cyberspace', *New Political Science*, nos 41–42, http://www.urbsoc.org/cyberpol/resnick.shtml.

17 Naturally, this distinction is an inexact one, as each type of Net politics will influence the others – tight government controls may affect the extent to which citizens can hope to effect social change, for example, while complicated internal politics within a group may in turn limit that group's effectiveness in its external orientation, and so on. But Resnick's taxonomy is still a useful one.

On politics within the Net see, for example, Nancy K. Baym (2000) *Tune In, Log On: Soaps, Fandom and Online Community*, Sage, Thousand Oaks, CA; and David Porter (ed.) (1997) *Internet Culture*, Routledge, London and New York. On politics which impact on the Net see, for example, Lawrence Lessig (1999) *Code: and Other Laws of Cyberspace*, Basic Books, New York; and Trevor Barr (2000) *newmedia.com.au*, Allen & Unwin, St Leonards, NSW.

Chapter 1

1 The Ruckus Society are at http://www.ruckus.org. ®™ark are at http://rtmark.com (we'll look at their work in detail in chapter 5). The Electrohippies are at http://www.fraw.org.uk/ehippies. The Electrohippies claim that more than 400,000 people took part in their virtual sit-in – see Dorothy Denning (2001) 'Cyberwarriors: Activists and Terrorists Turn to Cyberspace', *Harvard International Review*, vol. 23. no. 2, p. 71. We'll look in detail at the virtual sit-in tactic in chapter 6.

2 An extensive archive of Seattle coverage, including Deep Dish TV programs in Real Video format, is at http://seattle.indymedia.org.

A major irony of N30 and the subsequent global caravan of events was that its sophisticated Net use was designed to enable a very traditional tactic – the physical blockade of an institution. The protests were aimed at manifestations of informational capitalism; at network enterprise and global free trade agreements and corporate expansion. Yet the tactic in Seattle was one of the industrial age – a physical response to virtual power. For a comprehensive analysis of the features of informational capitalism which so-called 'anti-globalisation' protesters commonly oppose, see Manuel Castells (2000) *The Rise of the Network Society* (2nd edition), Blackwell, Oxford. For a critique of the physical blockade as a tactic for combating a globalised information economy, see Critical Art Ensemble (1995) *Electronic Civil Disobedience and Other Unpopular Ideas*, Autonomedia, New York. There are, of course, still things to be said for the blockade as a tactic: a large one can certainly call attention to an issue. But that attention can easily be neutralised by counter-spin – by the time of the S11 protests in Melbourne in 2000, for instance, Australian politicians were familiar with the tactics, and were armed with their soundbite that the protesters were 'against free speech'.

3 Paul Krugman (1999) 'Economic Luddites Rain on WTO's Parade', *The Australian*, 2 December, p. 13.

4 Thomas L. Friedman (1999) 'Protesters Fail to Separate Cause from Effect', *Sydney Morning Herald*, 2 December, p. 12.

5 Robert Garran (1999) 'Logic of Trade is Lost on Protestors', *The Australian*, 2 December, p. 7. It's striking how similar the media accounts of the anti-war demonstrations at the Pentagon in 1967 were to the media characterisation of the Seattle coalitions. In both cases, the fact that each represented diverse interests was presented by reporters as a contradiction, or as evidence of incoherence, rather than as evidence of a new emerging consensus or new coalitions. See, for example, *Time* magazine's coverage of the march on Washington of 21 October 1967: 'Protest: the Banners of Dissent' (1967) *Time*, vol. 90, no. 17, 27 October, pp. 9–15.

6 A position which reaches some kind of apotheosis in Thomas Friedman's 'Golden Arches Theory of Conflict Prevention', in which he extrapolates from the observation that 'No two countries that both had McDonald's had fought a war against each other since each got its McDonald's'. Thomas Friedman (1999) *The Lexus and the Olive Tree*, HarperCollins, London, pp. 195–217, quote from p. 195. An interesting counterpoint to this notion is *No Logo* author Naomi Klein's observations on the eve of the World Trade Organisation meeting in Qatar in the wake of the 11 September 2001 events: 'The battle lines leading up to next month's WTO negotiations in Qatar are: Trade equals freedom, antitrade equals fascism. Never mind that Osama bin Laden is a multimillionaire with a rather impressive global export network stretching from cash-crop agriculture to oil pipelines. And never mind that this fight will take place in Qatar, that bastion of liberty, which is refusing foreign visas for demonstrators but where bin Laden practically has his own TV show on the state-subsidized network Al-Jazeera. Naomi Klein (2001) 'Signs of the Times', *The Nation*, 21 October. Also archived at the nettime list at http://www.nettime.org.

7 In this regard, some kind of last word on the Seattle events must be that they've been turned into a Playstation game (titled 'State of Emergency').
8 Bush quoted in Francis Wheen (2000) 'It's Dumbo v Pinocchio', *The Guardian*, 25 October.
9 Figures from Paul Sheehan (2000) 'Leviathan Inc', *Sydney Morning Herald*, 15 January, p. 51.
10 On news and narrative see, for example, Elizabeth S. Bird and Robert W. Dardenne (1997) 'Myth, Chronicle and Story: Exploring the Narrative Qualities of News', in Dan Berkowitz (ed.) *Social Meanings of News*, Sage, Thousand Oaks, pp. 333–50; and Stuart Hall (1984) 'The Narrative Construction of Reality: an Interview', *Southern Review*, vol. 17, no. 1, pp. 3–17.
11 David Flynn (2000) 'The Earth Moved, But Freedom of Speech Stood Still at AOL', *Sydney Morning Herald*, 12 January, pp. 1, 6.
12 'Microsoft – Togetherness' (2000) *The Economist*, 22 January, p. 65.
13 The remaining two in the top 10 were the Australian federal government network, and the Google search engine, which, unlike most engines, eschews the portal trappings of free email, instant messaging, shopping promotions and 'content'. See *Jupiter Media Metrix* ratings http://au.jmm.com/top10.html.
14 The Hunger Site is at http://www.thehungersite.com. Its map is based on UN data showing that one death from hunger occurs every 3.6 seconds.
15 As of August 2001, The Hunger Site is under new ownership.
16 For a fuller development of this point see Sherman Young (2001) 'Brave New World of Digital', paper presented at *Digital Spaces, Public Places* conference, 16 November, University of Technology, Sydney. Transcript available at http://journalism.uts.edu.au/archive/digital/#transcripts.
17 See Katie Hafner and Matthew Lyon (1996) *Where Wizards Stay Up Late: the Origins of the Internet*, Touchstone, New York. Don't be put off by its terrible title – Hafner and Lyon's book is an authoritative history of the pre-web Net, based on extensive interviews with the actual originators of the ARPAnet. Several of those founders are among the collective authors of the following history, which is also recommended: Barry M. Leiner, Vinton G. Cerf, David D. Clark, Robert E. Kahn, Leonard Kleinrock, Daniel C. Lynch, Jon Postel, Larry G. Roberts and Stephen Wolff (2000) 'A Brief History of the Internet', http://www.isoc.org/internet/history/brief.shtml. Other histories are also collected at the same site (that of the Internet Society).

Two other excellent accounts of the Net's early development are John Naughton (1999) *A Brief History of the Future*, Weidenfeld & Nicolson, London; and Manuel Castells (2001) *The Internet Galaxy*, Oxford University Press, Oxford, chapter 1.

A detailed analysis of the origins of the packet-switching technologies which enabled the ARPAnet is Janet Abbate (1999) 'Cold War and White Heat: the Origins and Meanings of Packet Switching', in Donald MacKenzie and Judy Wajcman (eds) *The Social Shaping of Technology* (2nd edition), Open University Press, Buckingham, pp. 351–71. Packet-switching was theorised independently by both Paul Baran in the US and Donald Davies in the UK. An interview with Baran in which he tries to kill the nuclear-survival myth is Stewart Brand (2001)

'Founding Father', *Wired* 9.03, March, http://www.wired.com/wired/archive/9.03/baran.html. While Baran *was* working towards the Cold War communications survivalist ideal that has become the popular myth of the Net's origins, his research was used by ARPA towards very different ends. Davies, meanwhile, was motivated by Harold Wilson's 'white heat of technology' vision of revitalising the UK economy. It's not hard to see why the former offered a sexier basis for this founding Net myth, but – good story or not – it misrepresents the processes involved.

18 Tim Berners-Lee (1999) *Weaving the Web*, Orion Business Books, London, p. 36.

19 LabourStart is at http://www.labourstart.org. On the Internet and trade unions see Eric Lee (1996) *The Labour Movement and the Internet: the New Internationalism*, Pluto Press, London.

20 See, for example, John Aglionby (1998) '"Dozens Raped" During Days of Pillage', *Sydney Morning Herald*, 6 June, p. 22; and Seth Mydans (1998) 'Indonesians Report Widespread Rapes of Chinese in Riots', *New York Times*, 10 June.

21 World Huaren Federation, http://www.huaren.org.

22 See the Huaren editorial by Edward Liu: 'Internet Power, the Third Force, and the Global Huaren' (1998) http://www.huaren.org/editorial, 31 July.

23 The mission statements are at http://www.huaren.org/mission.

24 An archive of reports of the global protests, mixing established media with Huaren contributors, is at http://www.huaren.org/hottopic/id/090898-01.html. On the Beijing protest, see also Teresa Poole (1998) 'Chinese Protest at Mob-Rapes', *Independent International*, 19–25 August, p. 4.

25 'Alleged' victims in that there are question marks over the provenance of the pictures – *The Economist* wrote that the images were 'of outrages committed elsewhere at other times' (14 November 1998). While there's no doubt that such atrocities happened in the Indonesian events, there are suggestions that the actual photos on the Huaren site in fact document violence in East Timor – a suggestion which the Huaren spokespeople shrug off: 'Although we did try to check the source... we were unable to track it down,' says Joe Tan, adding, 'the real photos of Chinese victims could be just as bad, if not worse, from the stories of the victims. But again I cannot prove those photos were not Chinese victims.'

26 Huaren spokespeople emphasise that other groups of overseas Chinese were also heavily involved. Joe Tan identifies, for example, Indo Chaos, http://www.geocities.com/Pentagon/3233, as well as other groups with no website, who used email to co-ordinate the campaign. Chinese student organisations in Europe also played key roles. Tan also notes that many individuals and groups Huaren are not aware of would also have made substantial contributions to the international actions.

27 http://www.huaren.org/focus.

28 A thorough account of the Tiananmen media event can be found in McKenzie Wark (1994) *Virtual Geography*, Indiana University Press, Bloomington.

29 Constance Penley and Andrew Ross (eds) (1991) *Technoculture*, University of Minnesota Press, Minneapolis, pp. viii–ix. On the use of fax machines by the Tiananmen movement, see also Yan Ma (2000) 'Chinese Online Presence:

Tiananmen Square and Beyond', in Ann de Vaney, Stephen Gance and Yan Ma (eds) *Technology and Resistance: Digital Communications and New Coalitions Around the World*, Peter Lang, New York, pp. 139–51. Ma points to the irony that the communications technologies used by the students were those that had been promoted by the regime itself in the interests of modernisation. Ma's article includes an informal survey of newspaper editors which supports Penley and Ross's contention that the use of fax machines in the 1989 events was primarily to *receive* information from overseas sources rather than to *transmit* information from the scene.

30 Arjun Appadurai (1996) *Modernity at Large: Cultural Dimensions of Globalization*, University of Minnesota Press, Minneapolis, p. 4.

31 Brian Whitaker and Patrick Barkham (2000) 'Sites Caught in Saudis' Web', *The Guardian Weekly*, 18–24 May, p. 21.

32 A participant's account of this is Shahidul Alam (1996) 'On-Line Lifeline', *New Internationalist*, no. 286, p. 14. Archived at http://www.oneworld.org/ni.

33 On censorship and the Malaysian Multimedia Super Corridor, see Peter Alford (2000) 'Mahathir Can't Silence his Cyber Critics', *The Australian* Media supplement, 30 March, p. 8.

34 See, for example, Louise Williams (1998) 'May Riots: Facts Not Enough for Military Chief', *Sydney Morning Herald*, 5 November, p. 12; and Don Greenlees (1998) 'Rape's Terror Reaches Beyond Physical Scars', *The Australian*, 4 November, p. 9. In April 1999, the UN Special Rapporteur on violence against women, Radhika Coomaraswamy, reported that she had found enough evidence on her visit to Indonesia in November 1998 to convince her that there had been widespread rapes of ethnic Chinese women and that the security forces had at best done nothing, at worst colluded. She also described how, following the riots, victims had been dissuaded from reporting the offences by letters including photographs of their own rape, along with threats that these would be widely circulated if they spoke out.

35 An interesting discussion of the limitations of news groups as public forum is Joseph Knapp (1997) 'Essayistic Messages: Internet Newsgroups as an Electronic Public Sphere', in David Porter (ed.) *Internet Culture*, Routledge, London and New York, pp. 181–97.

36 The Nuremberg Files is at http://www.christiangallery.com/atrocity/index.html.

37 See Ian Lucas (1998) *OutRage! An Oral History*, Cassell, London.

38 Despite its closure in 1999, the site continued to reappear from time to time on different servers, and could usually be found by searching for it on a good search engine, such as Google. A useful archive of media coverage of the site is at http://www.xs4all.nl/~kspaink/nuremberg. On the broader context of anti-abortion activism, see Patricia Baird-Windle and Eleanor J. Bader (2001) *Targets of Hatred: Anti-Abortion Terrorism*, Palgrave, Basingstoke.

39 See, for example, Duncan Campbell (2001) 'Seven Doctors Murdered, Now US Judges Rule In Favour of Abortion Hit List', *The Guardian*, 30 March.

40 See, for example, Helen Trinca (1999) 'Unions Take a Byte of the Boss', *Sydney Morning Herald*, 18 February, p. 10. Examples of this can also be found through the online archives of Eric Lee's LabourStart.

41 This 'House and Garden competition' is at http://www.lockstockandbarrel.org/Documents/HOUSE-GARDEN.htm.

42 Marshall McLuhan and Quentin Fiore (1967) *The Medium is the Massage*, Hardwired, San Francisco, pp. 74–75.

43 I am, of course, focusing here only on the traditional tactics of *non-violent* activism. An exhaustive analysis of these is Gene Sharp (1973) *The Politics of Nonviolent Action* (in 3 volumes), Porter Sargent, Boston. See also the excellent history of civil disobedience and non-violent resistance, from Gandhi to Solidarity and many other case studies, in Peter Ackerman and Jack Du Vall (2000) *A Force More Powerful: a Century of Nonviolent Conflict*, Palgrave, New York and Basingstoke. Also useful are Saul D. Alinsky (1971) *Rules For Radicals: a Pragmatic Primer for Realistic Radicals*, Vintage Books, New York; and Phil Thornton, Liam Phelan and Bill McKeown (1997) *I Protest!*, Pluto Press, Sydney. Mark Edelman Boren (2001) *Student Resistance: a History of the Unruly Subject*, Routledge, New York and London, offers some detailed accounts – if almost no analysis – of activist tactics in student campaigns, both non-violent and violent.

On more recent developments in activist tactics and their changing context (some of which are examined in chapter 6), see John Arquilla and David Ronfeldt (1998) *The Zapatista Social Netwar in Mexico*, RAND Corporation report, available online at http://www.rand.org/publications/MR/MR994.

There are many how-to guides online for nascent Net activists. See, for example, Audrie Krause, Michael Stein, Judi Clark, Theresa Chen, Jasmine Li, Josh Dimon, Jennifer Kanouse and Jill Herschman (2001) *NetAction's Virtual Activist Training Guide 2.0* http://netaction.org/training; Phil Agre (1999) 'Designing Effective Action Alerts for the Internet', http://dlis.gseis.ucla.edu/people/pagre/alerts.html; and *ACT UP Civil Disobedience Manual* (no date) http://www.actupny.org/documents/CDdocuments/CDindex.html. The Institute for Global Communications, a key hub for Internet activism, offers archived advocacy tips at http://www.igc.org/igc/gateway/index.html. UK journalist and academic George Monbiot (no date) authored *An Activist's Guide to Exploiting the Media*, http://www.urban75.com/Action/media.html. See also Protest.Net's *Activists Handbook* (no date) http://yada.protest.net/activists_handbook and the 'Get Active' mini-site within Greenpeace Australia http://www.greenpeace.org.au/getactive/index.html.

44 See Matthew Arnison (2001) 'Crazy Ideas for Webcasting', http://www.cat.org.au/cat/webcast.html.

45 Alinsky, *Rules For Radicals*, p. 41. I take Alinsky's remark to be about the *specific* tactics used by Gandhi: it would be wrong to suggest that *all* non-violent tactics were useless against the Nazis. Examples of non-violent resistance in Nazi-occupied Europe can be found in Sharp, *The Politics of Nonviolent Action*, pp. 87–90; and Ackerman and Du Vall, *A Force More Powerful*, pp. 207–39.

Chapter 2

1 'New Remote Control Can Be Operated By Remote', in Scott Dikkers and Robert Siegel (eds) (2000) *The Onion's Finest News Reporting*, Boxtree, London, pp. 84–85. *The Onion* is online at http://www.theonion.com.

2 Eno, quoted in his interview with Kevin Kelly (1995) 'Gossip is Philosophy', *Wired* 3.05, May, pp. 146–51 and 204–209, online at http://www.wired.com/wired/archive/3.05/eno.html.

3 On loose uses of 'interactive': for example, in an otherwise excellent book, sociologist Manuel Castells treats the meaning of 'interactive' as a given, and doesn't examine what conflicting or multiple meanings the term might have. Castells runs together what I would term Version 1.0 uses (democratising, enabling) with Version 2.0 uses (the emphasis on consumer 'choice') as though they were no different from each other. Castells's approach is symptomatic of a wider problem: his section on 'the interactive society' is, as he says, primarily a literature review of writings on computer-mediated communication. The fact that this leaves him with little to say on interactivity is a function of the fact that few writers on the topic unpack what 'interactivity' really means. See Manuel Castells (1996) *The Rise of the Network Society*, Blackwell, Oxford, chapter 5.

For more examples of the loose uses of the term, see, for instance, the feature issue 'Interactive' of online media and cultural studies journal *M/C Reviews*. While several of its twelve articles acknowledge the problems of the word 'interactivity', none attempts to untangle its often contradictory meanings. If the promise of interactivity as *production* often conflicts with its applications as *consumption*, the editors of this issue are in on the act: 'Interactivity, in its basic form', they write, 'refers to some form of *consumer* involvement, usually in the form of a direct action, *within the product they consume*. It proffers increased information, access, power and even control for *consumers*' [emphasis added]. See Kelly McWilliam and Kate Douglas (2001) 'Editorial: Your Say – Interactivity in Contemporary Media', *M/C Reviews* http://www.media-culture.org.au/reviews/features/interactive/editorial.html.

4 Jens F. Jensen (1999) '"Interactivity" – Tracking a New Concept in Media and Communication Studies', in Paul A. Mayer (ed.) *Computer Media and Communication: a Reader*, Oxford University Press (OUP), pp. 160–87. It's worth noting that in my *Economist* example, nothing is unique to the Net: I could subscribe to the hard copy of the magazine and of course choose when to read it and which stories to read. I could allow the magazine to collect data about me by responding to readership questionnaires. And, obviously enough, I could have my email flame war on the phone or in person.

Also of real interest in this area is Edward J. Downes and Sally J. McMillan (2000) 'Defining Interactivity: a Qualitative Identification of Key Dimensions', *New Media & Society*, vol. 2, no. 2, pp. 157–79. Downes and McMillan suggest six key dimensions of interactivity: the first three focus on messages; the second three focus on participants. These are: the direction of communication flows; time flexibility, such as the potential of email and newsgroups for asynchronous communication; the sense of place created by cyberspace communications; the degree of control afforded to the participant; the need and potential for participants to respond to messages and receive responses themselves; and the perceived goal of a given communication. While this argument, like Jensen's, is very important in that it seriously attempts to define a concept which is so often taken for granted, its separation of 'message' and 'participant' dimensions of interactivity ends up erecting a barrier where it could instead be taking one down.

See also Erkki Huhtamo (1999) 'From Cybernation to Interaction: a Contribution to an Archaeology of Interactivity', in Peter Lunenfeld (ed.) *The Digital Dialectic: New Essays on New Media,* MIT Press, Cambridge, MA, pp. 96–110. Huhtamo traces some of the discourses surrounding interactivity to those of 'automation' in the 1950s and early 1960s, pointing to the tension between production and consumption which *The Onion*'s remote control parody nails. See also the virtual sex parody FuckU-FuckMe™ by digital artist Alexej Shulgin at http://www.fufme.com. This site purports to be a product catalogue for computer sex hardware and software interfaces – 'the most complete remote sex solution for the Internet and corporate intranet'. The barb in the satire is aimed at Version 2.0-style interactivity: at a world view in which sex, like politics, just isn't remote enough, and can be improved upon by a programmed 'intuitive interface' available for Windows 95, 98 and NT users. As a product which 'allows you to entirely concentrate on remote communication', FuckU-FuckMe may be the last word on remote-control-style interactivity.

5 Kelly, 'Gossip is Philosophy'.
6 Tim Berners-Lee (1999) *Weaving the Web*, Orion Business Books, London, pp. 182–83.
7 We should note, though, that consultational interactivity was also seen as important, as it was argued that we would be able to access otherwise hard-to-find information, information which could make a difference to our civic participation and cultural life. Howard Rheingold, for instance, looked back to the founding principles of representative democracy to remind us of the argument that such governance implies informed consent – and that this consent could be more fully informed as new channels of information were developed to balance and broaden those of the commercialised media. See Rheingold, *The Virtual Community*, pp. 13–14.
8 On the concept of virtual community, Saul Alinsky (1969) *Reveille For Radicals*, Vintage Books, New York, should be required reading for anyone thinking that tech tools alone will enable them to start a 'community'. The idea that 'if you build it, they will come', isn't enough, as Alinsky's accounts of building community organisations from the bottom up demonstrate. See also Phil Agre (1999) 'Avoiding Heat Death on the Internet', in Josephine Bosma, Pauline Van Mourik Broekman, Ted Byfield, Matthew Fuller, Geert Lovink, Diana McCarty, Pit Schultz, Felix Stalder, McKenzie Wark and Faith Wilding (eds) *README!: ASCII Culture and the Revenge of Knowledge*, Autonomedia, New York, pp. 343–56.
9 The ARPA experiments which became the foundation of the Net were focused on interactive computing, but in the sense of direct human–machine interaction, dispensing with the clumsiness of piles of printouts or punch cards.

The conversational dimension of the Net is a by-product of resource-sharing technologies. Email was one of the first applications of resource-sharing networks – first as users compared notes on what they were working on, but also for jokes, personal notes and the building of relationships. Rheingold describes this as 'automatically building in the potential for a community' (*The Virtual Community*, p. 72). What's interesting about this is that his concept of community is predicated upon the potential for conversational interactivity – precisely the quality least prominent in the GUI-era of the web – yet the community metaphor has become increasingly commonplace even as the web becomes more central to the online environment. The possibility of virtual community was discussed as early as 1968 in an essay by two ARPA pioneers: see J.C.R. Licklider and Robert W. Taylor (1999) 'The Computer as Communication Device', in Paul A. Mayer (ed.) *Computer Media and Communication: a Reader*, OUP, pp. 97–110.

10 See Barlow's home page: http://www.eff.org/~barlow/barlow.html.

11 John Perry Barlow, Sven Birkerts, Kevin Kelly and Mark Slouka (1995) 'What Are We Doing On-Line?' *Harper's Magazine*, August, pp. 35–46.

12 John Perry Barlow (1996) 'A Declaration of the Independence of Cyberspace', http://www.eff.org/pub/Publications/John_Perry_Barlow/barlow_0296.declaration.

13 Mitch Kapor (1993) 'Where is the Digital Highway Really Heading?', *Wired* 1.03, http://www.wired.com/wired/archive/1.03/kapor.on.nii.html.

14 The WELL is on the web at http://www.well.com. An interesting account of its development is Katie Hafner (1997) 'The Epic Saga of the Well', *Wired* 5.05, http://www.wired.com/wired/archive/5.05/ff_well.html. On the changing – perhaps reduced – significance of such communities in the face of a more diffuse Internet, see Manuel Castells (2001) *The Internet Galaxy: Reflections on the Internet, Business and Society*, OUP, Oxford, chapter 4.

15 Rheingold, *The Virtual Community*, p. 6.

16 Ibid., p. 4.

17 On the public sphere, see Jurgen Habermas (1974) 'The Public Sphere: an Encyclopedia Article', *New German Critique*, vol. 1, no. 3, pp. 49–55 and (1989) *The Structural Transformation of the Public Sphere*, MIT Press, Cambridge, MA. On the limitations of applying the public sphere model to the Net, see Lincoln Dahlberg (1998) 'Cyberspace and the Public Sphere: Exploring the Democratic Potential of the Net', *Convergence*, vol. 4, no. 1, pp. 70–84.

18 What I see as Rheingold's pragmatic optimism has also been denounced as a kind of cyberhype, though. Academics Richard Barbrook and Andy Cameron, for example, label Rheingold a 'guru' of what they term the 'Californian Ideology' – a world view which 'promiscuously combines the free-wheeling spirit of the hippies and the entrepreneurial zeal of the yuppies'. This Californian Ideology is described as an emerging global orthodoxy on social and technological processes, in which 'technology is once again being used to reinforce the difference between the masters and the slaves'. See Richard Barbrook and Andy Cameron (1995) 'The Californian Ideology', http://www.wmin.ac.uk/media/HRC/ci/calif5.html.

19 See McKenzie Wark (1997) 'Cyberhype: the Packaging of Cyberpunk', in Ashley Crawford and Ray Edgar (eds) *Transit Lounge*, Craftsman House, Sydney, pp. 154–57 and, in the same volume, 'Infohype', pp. 144–49.
20 Darren Tofts and Murray McKeich (1998) *Memory Trade: a Prehistory of Cyberculture*, Interface, Sydney, p. 19.
21 Ibid., p. 19.
22 Paul Frissen (1997) 'The Virtual State: Postmodernisation, Informatisation and Public Administration', in Brian D. Loader (ed.) *The Governance of Cyberspace*, Routledge, London and New York, pp. 111–25, quote from p. 121.
23 Malcolm Waters (1995) *Globalization*, Routledge, London and New York, pp. 163–64.
24 Cited in Howard Besser (1995) 'From Internet to Information Superhighway', in James Brook and Iain A. Boal (eds) *Resisting the Virtual Life*, City Lights, San Francisco, pp. 59–70, quote from p. 60.
25 Langdon Winner (1986) *The Whale and the Reactor*, University of Chicago Press, Chicago and London, p. 20.
26 Version 2.0 cyberhype may be part of the wider ideological position that cultural critic Thomas Frank labels 'market populism' – a position that depicts the workings of the market as the expression of the popular will. Frank sees this position expressed across the full spectrum of 1990s US culture, from popular business literature to politicians, from *Wired* magazine to humanities departments: Thomas Frank (2000) *One Market Under God: Extreme Capitalism, Market Populism, and the End of Economic Democracy*, Doubleday, New York. On the real-life consequences for people who have to work in this market, with its Version 2.0 buzzwords of 'flexibility', 'teamwork' and 'choice', see Richard Sennett (1998) *The Corrosion of Character: the Personal Consequences of Work in the New Capitalism*, Norton, New York and London. And for an activist polemic urging the appropriation of such language and its ideas, see Geert Lovink and Florian Schneider (2001) 'New Rules of the New Actonomy', http://new.actonomy.org. 'Remember,' they write, 'that activism and entrepreneurial spirit have a lot in common.' 'Flexible' response and cellular 'team' structure are important components of tactical media groups such as ®™ark, Critical Art Ensemble and the Electronic Disturbance Theater projects, discussed in later chapters of this book. In the case of ®™ark, reclaiming corporate language is a key part of their work; their use of their registered corporate status also points to a key tactical idea – engaging with their opponent on their terms, appropriating resources from wherever they can, even from business literature.
27 Esther Dyson, Alvin Toffler, George Gilder and George Keyworth (1994) 'Cyberspace and the American Dream: a Magna Carta for the Knowledge Age', http://www.pff.org/position.html.
28 Nicholas Negroponte (1995) *Being Digital*, Hodder & Stoughton, Rydalmere, NSW, p. 6.
29 Negroponte, quoted in 'Barbie Will Hog the Net' (2000) *The Australian*, 13 June, p. 41. The Media Lab's sponsors are listed at http://www.media.mit.edu/Sponsors.

For other examples of Version 2.0 cyberhype see George Gilder (1994) *Life After Television*, Norton, New York. See also pretty much any issue of *Wired* (if you have to pick just one, go for the fifth anniversary issue: 6.01, January 1998). There is much good writing in *Wired*, and I've cited some of it in this book. But I do think we need to read it with our hype filters turned up to 10.

30 Vivian Sobchack (1994) 'New Age Mutant Ninja Hackers: Reading *Mondo 2000*', in Mark Dery (ed.) *Flame Wars: the Discourse of Cyberculture*, Duke University Press, Durham, pp. 11–28; Mark Dery (1996) *Escape Velocity: Cyberculture at the End of the Century*, Grove Press, New York.

31 Dery, *Escape Velocity*, p. 40.

32 Wark 'Cyberhype: the Packaging of Cyberpunk', p. 157.

Also of interest here is Ross Harley (1996) 'That's Interaction: Audience Participation in Entertainment Monopolies', *Convergence*, vol. 2, no. 1, pp. 103–23. While it doesn't address the Net directly, Harley's essay can be read as a critique – or a prehistory – of cyberhype: he notes how new technologies are sold to us on their revolutionary potential – what he terms their 'discourse of enablement' – while tracing how this potential is sublimated to the demands of industry. What his essay has in common with those critiques which do explicitly address the cyberhype of the Net is that it questions the commercial co-opting of radical potential – and the co-opting of the *language* of that potential.

33 Arthur Kroker (1996) 'Virtual Capitalism', in Stanley Aronowitz, Barbara Martinson and Michael Menser, with Jennifer Rich (eds) *Technoscience and Cyberculture*, Routledge, London and New York, pp. 167–79, quote from p. 170.

34 See, for example, the subtitles of these well-known books: Howard Rheingold (1994) *The Virtual Community: Homesteading on the Electronic Frontier* (the subtitle varies across editions); Katie Hafner and John Markoff (1991) *Cyberpunk: Outlaws and Hackers on the Computer Frontier*, Corgi, London; Bruce Sterling (1992) *The Hacker Crackdown: Law and Disorder on the Electronic Frontier*, Viking, London.

35 Douglas Rushkoff (1994) *Cyberia: Life in the Trenches of Hyperspace*, Harper, San Francisco.

36 Quoted in Edward S. Herman and Robert W. McChesney (1997) *The Global Media: the New Missionaries of Corporate Capitalism*, Cassell, London, p. 126.

37 Anthony Browne (1998) 'Posse Heads For Tax Fight at www.ok.corral', *Guardian Weekly*, 7 June, p. 19.

38 Ziauddin Sardar (1996) 'alt.civilizations.faq: Cyberspace as the Darker Side of the West', in Ziauddin Sardar and Jerome Ravetz (eds) *Cyberfutures*, Pluto Press, London, pp. 14–31, quote from p. 15. For other critiques on aspects of cyberhype see (inter alia) Kevin Robins (1995) 'Cyberspace and the World We Live In', in Mike Featherstone and Roger Burrows (eds) *Cyberspace/Cyberbodies/Cyberpunk: Cultures of Technological Embodiment*, Sage, London, pp. 135–55; Peter Golding (1999) 'Worldwide Wedge: Division and Contradiction in the Global Information Infrastructure', in Paul Marris and Sue Thornham (eds) *Media Studies: a Reader* (2nd edition), Edinburgh University Press, pp. 802–15; and the essays collected in James Brook and Iain A. Boal, *Resisting the Virtual Life*.

39 Rheingold, *The Virtual Community*, p. 2.

40 Julian Dibbell (1993) 'A Rape in Cyberspace', http://www.levity.com/julian/bungle_vv.html.
41 Sherry Turkle (1995) *Life on the Screen: Identity in the Age of the Internet*, Simon & Schuster, New York.
42 See the Electronic Frontier Foundation site: http://www.eff.org. The background to the formation of the EFF is discussed in both Rheingold, *The Virtual Community* and Bruce Sterling (1994) *The Hacker Crackdown*, Penguin, Harmondsworth. The full text of Sterling's book is online at a number of locations, including http://www.eff.org/Publications/Bruce_Sterling/Hacker_Crackdown. Many hackers use the term 'cracker' to distinguish those who make illegal use of stolen data, or who maliciously damage information or systems, from those with less malign motivations.
43 See the Electronic Frontier Foundation archive on the case at http://www.eff.org/pub/Legal/Cases/Canter_Siegel.
44 See, for example, Peter Kollock (1999) 'The Economies of Online Cooperation', in Marc A. Smith and Peter Kollock (eds) *Communities in Cyberspace*, Routledge, London and New York, pp. 220–39.
45 See Peter H. Lewis (1996) 'Protest, Cyberspace-Style, for New Law', *New York Times*, 8 February, p. A16.
46 For a detailed analysis of this campaign, see Laura J. Gurak (1997) *Persuasion and Privacy in Cyberspace: the Online Protests over Lotus MarketPlace and the Clipper Chip*, Yale University Press, New Haven.
47 Langdon Winner (1991) 'A Victory for Computer Populism', *Technology Review*, vol. 94, no. 4, p. 66.
48 John R. Wilke (1990) 'Lotus Product Spurs Fears About Privacy', *Wall Street Journal*, 13 November, pp. B1 and B5.
49 Minor parties, unable to count on media attention, tend to make more use of the Net, as we'll see below on One Nation. The Australian Greens, for instance, run several mailing lists, http://www.greens.org.au. My comments here relate to the following political party websites: in Australia, the Democrats, http://www.democrats.org.au; the Australian Labor Party, http://www.alp.org.au; the Liberals http://www.liberal.org.au; the National Party, http://www.nationalparty.org. In the UK, the Conservatives, http://www.conservatives.com/home.cfm; the Labour Party, http://www.labour.org.uk; the Liberal Democrats, http://www.libdems.org.uk; the Scottish Nationalist Party, http://www.snp.org.uk. In the US, the Democratic National Committee, http://www.democrats.org, and the Republican National Committee, http://www.rnc.org; and also the 2000 Presidential campaign sites of Al Gore, http://www.algore2000.com (now no longer active) and George W. Bush, http://www.georgewbush.com.
50 Quoted in Selina Mitchell (2001) 'Labor Keeps it Simple to Get Voters Into its Net', *The Australian*, 22 March, p. 2.
51 Prime Minister of Australia, http://www.pm.gov.au.
52 Sinn Fein, http://www.sinnfein.ie.
53 The 'digital homeless' quote is from an interview with Negroponte on ABC television titled 'Digital Nostradamus', *Lateline*, 19 March 1996.

54 Juliet Roper (1997) 'New Zealand Political Parties Online: the World Wide Web as a Tool for Democratization or for Political Marketing?', in *New Political Science*, nos. 41–42, http://www.urbsoc.org/cyberpol/roper.shtml, section 18.

55 Paul Nixon and Hans Johansson (1999) 'Transparency Through Technology: the Internet and Political Parties', in Barry N. Hague and Brian D. Loader (eds) *Digital Democracy: Discourse and Decision Making in the Information Age*, Routledge, London and New York, pp. 135–53, quote from p. 148.

56 Wayne Rash (1997) *Politics on the Nets: Wiring the Political Process*, W.H. Freeman, New York, p. 17.

57 Stephen Coleman (1999) 'The New Media and Democratic Politics', *New Media & Society*, vol. 1, no. 1, pp. 67–74, quote from p. 70.

58 See, for example, Jonathan Watts (2000) 'Putin Tells World Leaders to Log On', *The Observer*, 23 July.

59 For a detailed analysis of this, see Joshua Meyrowitz (1985) *No Sense of Place*, OUP, New York, particularly chapter 14. While it's possible to read Meyrowitz as a technological determinist, it's also important to note the extent to which his argument addresses the equally complex question of how the communications environment is shaped by political uses.

60 See Coleman, 'The New Media and Democratic Politics', p. 68, and Manuel Castells (1997) *The Power of Identity*, Blackwell, Oxford, pp. 317–23.

61 Meyrowitz, *No Sense of Place*.

62 Alan Travis (2001) 'Campaign Blamed for Low Poll', *The Guardian*, 4 July.

63 Castells, *The Power of Identity*, p. 312.

64 Dick Morris (1999) *Vote.com*, Renaissance Books, Los Angeles, p. ix. On the Fourth Estate, see Julianne Schultz (1998) *Reviving the Fourth Estate*, Cambridge University Press, Cambridge, UK.

65 A useful overview of critical thinking on polls is Justin Lewis (1999) 'The Opinion Poll as Cultural Form', *International Journal of Cultural Studies*, vol. 2, no. 2, pp. 199–221.

66 *Future Exchange* http://www.abc.net.au/future.

67 The distinction between opinions and emotions is made with regard to television news by Neil Postman (1985) *Amusing Ourselves to Death*, Penguin, New York.

68 Public Debate, http://www.publicdebate.com.au.

69 Morris, *Vote.com*, p. 184.

70 Morris, *Vote.com*, p. 29.

71 Quoted in David Humphries (2001) 'One Nation Facing Poll Disaster, Says Oldfield', *Sydney Morning Herald*, 26 January, p. 6.

72 Margo Kingston (1999) *Off The Rails: the Pauline Hanson Trip*, Allen & Unwin, St Leonards, NSW, p. 42.

73 Kingston was not the only journalist to recognise and grapple with this dilemma. See, for example, Leisa Scott (1998) 'Strain of Costing Policies is Adding up for Hanson', *The Australian*, 30 September, p. 9.

74 On resistance identity see Castells, *The Power of Identity*.

75 An interesting analysis in this vein comes from Don Fletcher and Rosemary Whip, who see One Nation as symptomatic of a broader failure in political

leadership. Fletcher and Whip point to the increasing 'professionalisation' of politics, in several senses: the rise of the career politician; the increasing number of politicians who come from the professions and hence have little in common with many other demographics; and the entrenchment of slick, spin-doctored campaigns. See Don Fletcher and Rosemary Whip (2000) 'One Nation and the Failure of Political Leadership', in Ian Ward, Michael Leech and Geoffrey Stokes (eds) *The Rise and Fall of One Nation*, University of Queensland Press, St Lucia, pp. 73–85.

76 The 1998 campaign site is now accessible only on a paid-subscription basis, although it is partly archived at the National Library of Australia's Pandora project, http://pandora.nla.gov.au.

In the 2001 federal election not only were One Nation a significantly reduced force and media non-event, but their website was skeletal (a consequence, perhaps, of having purged their webmaster Scott Balson from the party in 1999). So what I have to say about them in this section applies only to the 1998 campaign, which is still of historical interest.

77 See the introduction to Ian Ward, Michael Leech and Geoffrey Stokes (eds) (2000) *The Rise and Fall of One Nation*, a collection of that conference's papers. Balson's own account of the conference offers examples of several of One Nation's key tactics, including personal attacks on critics, the labelling of all disagreement as 'bias', and the catch-all use of the word 'elite' as a smear. See Scott Balson (1998) 'Where Prize Turkeys Gather', 15 November, http://www.gwb.com.au/gwb/news/onenation/uqforum.

78 http://www.gwb.com.au/gwb/news/onenation/federal/final.

79 Rash, *Politics on the Nets: Wiring the Political Process*, p. 38.

80 Castells, *The Power of Identity*, p. 351.

81 Analysing such online discussions presents significant problems, which we need to acknowledge. One complicating factor is that participants often devote a lot of energy to policing the boundaries of their group: Michele Tepper, for example, has drawn attention to some of the ways in which group members act to police their own boundaries: the phenomenon of 'trolling' is one instance – joke posts which only 'insiders' will recognise; those who respond expose themselves as outsiders, allowing the group to reinforce their collective identity by exclusion. A similar phenomenon is the posting of 'flamebait': inflammatory messages calculated to provoke outrage. Such practices make it difficult to adopt the interpretive perspective of a disinterested observer: there is, in short, a certain degree of risk in taking posts at face value, given that the complex dynamics which can evolve within a group generate ambiguities, subtleties and hidden agendas which are not apparent to the observer. Indeed, as with trolling, they are *not supposed* to be apparent to the observer. See Michele Tepper (1997) 'Usenet Communities and the Cultural Politics of Information', in David Porter (ed.) *Internet Culture*, Routledge, London and New York, pp. 39–54.

82 Brian A. Connery (1997) 'IMHO: Authority and Egalitarian Rhetoric in the Virtual Coffeehouse', in David Porter, *Internet Culture*, pp. 161–79, quote from p. 174.

83 David Resnick (1997) 'Politics on the Internet: the Normalization of Cyberspace', *New Political Science*, nos. 41–42, http://www.urbsoc.org/cyberpol/resnick.shtml, section 15.

84 James A. Knapp (1997) 'Essayistic Messages: Internet Newsgroups as an Electronic Public Sphere', in David Porter, *Internet Culture*, pp. 181–97.

85 Castells, *The Power of Identity*, p. 351.

86 For an analysis of this question with regard to Usenet newsgroups, see Knapp, 'Essayistic Messages'.

87 McKenzie Wark (1998) 'Net.politics and the Virtual Republic', *Mute*, issue 11, http://www.metamute.com/issue11/wark.htm, accessed 21 October 1998 (no longer available).

Chapter 3

1 John Downing (1995) 'Alternative Media and the Boston Tea Party', in John Downing, Ali Mohammadi and Annabelle Sreberny-Mohammadi (eds) *Questioning The Media*, Sage, Thousand Oaks, CA, pp. 238–52, quote from p. 240. By 2001, though, Downing was to write that 'to speak simply of alternative media is almost oxymoronic. Everything, at some point, is alternative to something else.' Instead, Downing now proposes the term 'radical alternative media'. My own view is that 'alternative' remains the best of an unsatisfactory bunch of terms: 'radical', for instance, connotes, to many people, issues and movements broadly of the Left, whereas the term 'alternative', as I use it here, allows space to examine the Internet use of groups with a more right-wing agenda (such as the Nuremberg Files or One Nation, for example). 'Independent', similarly, often connotes *economic* independence, skewing our reading towards questions of political economy alone. See John D.H. Downing with Tamara Villarreal Ford, Geneve Gil and Laura Stein (2001) *Radical Media: Rebellious Communication and Social Movements*, Sage, Thousand Oaks, CA, pp. ix–xi (the Downing quote above is from p. ix). On the difficulties of defining 'alternative', see also Chris Atton (2002) *Alternative Media*, Sage, London. On the problems with the term 'independent', see, for instance, the essays collected in Jim Hillier (ed.) (2001) *American Independent Cinema: a Sight and Sound Reader*, British Film Institute, London.

Given the long tradition in media studies of critical opposition to corporate control, there are fewer studies of alternative media than might be expected, and those studies which exist do not always lend themselves readily to an analysis of the Internet. The question of alternative media has typically been phrased in terms of ownership and control, issues which, while still important, are reframed by the Internet. Downing's work includes a number of other important contributions, including (1989) 'Computers for Political Change: Peacenet and Public Data Access', *Journal of Communication*, vol. 39, no. 3, pp. 154–62; and (1988) 'The Alternative Public Realm: the Organization of the 1980s Anti-Nuclear Press in West Germany and Britain', *Media, Culture and Society*, vol. 10, no. 2, pp. 163–81. The 1984 edition of his *Radical Media* (South End Press, Boston), while dated, still contains much of historical interest.

Other examples of interest are: Jesse Drew (1995) 'Media Activism and Radical Democracy', in James Brook and Iain A. Boal (eds) *Resisting the Virtual Life*, City Lights, San Francisco, pp. 71–83. Drew's discussion of Paper Tiger TV and the Deep Dish Satellite Network focuses on the potential of camcorders and satellites to circumvent the gatekeepers of corporate broadcast networks. On alternative television, see also DeeDee Halleck (1984) 'Paper Tiger Television: Smashing the Myths of the Information Industry Every Week on Public Access Cable', *Media, Culture and Society*, vol. 6, no. 3, pp. 313–18, and Douglas Kellner (1990) *Television and the Crisis of Democracy*, Westview Press, Boulder, CO. An insider's account of the development of alternative video magazine *Undercurrents* is Thomas Harding (1998) '*Viva Camcordistas!* Video Activism and the Protest Movement', in George McKay (ed.) *DiY Culture: Party & Protest in Nineties Britain*, Verso, London, pp. 79–99. Barbara Trent's account of the making of her Academy Award-winning documentary *The Panama Deception* details the economic and political obstacles for dissident filmmakers in the US: Barbara Trent (1998) 'Media in a Capitalist Culture', in Fredric Jameson and Masao Miyoshi (eds) *The Cultures of Globalization*, Duke University Press, Durham and London, pp. 230–46. On questions of access, production and distribution with regard to radio, see the polemical pamphlet by Greg Ruggiero (1999) *Microradio & Democracy: (Low) Power to the People*, Seven Stories Press, New York. On print, see the analyses of independent newspapers and magazines by Chris Atton (1999) 'A Reassessment of the Alternative Press', *Media, Culture and Society*, vol. 21, no. 1, pp. 51–76 and Susan Forde (1998) 'Monitoring the Establishment: the Development of the Alternative Press in Australia', *Media International Australia*, no. 87, pp. 114–33. A useful historical context is the account of the early radical press in the UK, in James Curran and Jean Seaton (1981) *Power Without Responsibility*, Fontana, London. More wide-ranging but of primarily historical interest is Armand Mattelart and Seth Siegelaub (eds) (1983) *Communication and Class Struggle: 2. Liberation, Socialism*, International General, New York.

2 On this point see Thomas Frank (1997) *The Conquest of Cool: Business Culture, Counterculture, and the Rise of Hip Consumerism*, University of Chicago Press, Chicago and London. Frank's book is a rich cultural history of the advertising and fashion industries in the US in the 1960s, with much to say about those industries' appropriation of 'hip' culture (although his stated decision to not consider the cultural *reception* of this 'hip' imagery does mean that the book occasionally reads as though it's suggesting that cultural change happens *only* through advertising).

3 This coverage is archived at http://www.guardian.co.uk/gmdebate.

4 David S. Bennahum (1997) 'The Internet Revolution', *Wired* 5.04, April, http://www.wired.com/wired/archive/5.04/ff_belgrad.html.

5 Matt Drudge is at http://www.drudgereport.com.

6 DeathNET is at http://www.rights.org/deathnet.

7 North American Man/Boy Love Association, http://www.nambla.de. NAMBLA were the subject of a classic episode of *South Park:* on the relationships between satire and activism, see chapter 5.

8 On the relationships between B92's music policy and their oppositional politics, see Matthew Collin (2001) *This Is Serbia Calling: Rock'n'Roll Radio and Belgrade's Underground Resistance*, Serpent's Tail, London.

9 Collin, *This Is Serbia Calling*, p. 57.

10 The Digital City is at http://www.dds.nl – in December 1999, it adopted a more corporate structure. On this shift towards a more Version 2.0 model, see Geert Lovink and Patrice Riemens (2000) 'Amsterdam Public Digital Culture 2000', http://heise.de/tp/english/inhalt/co/6972/1.html; see also Manuel Castells (2001) *The Internet Galaxy*, Oxford University Press, Oxford, pp. 146–55.

11 A selection of essays and other writing by Geert Lovink can be found at http://thing.desk.nl/bilwet. XS4ALL is at http://www.xs4all.nl. The nettime mailing list is archived on the web and is a useful resource for further reading on B92: http://www.nettime.org. A compilation of posts to the list is Josephine Bosma, Pauline Van Mourik Broekman, Ted Byfield, Matthew Fuller, Geert Lovink, Diana McCarty, Pit Schultz, Felix Stalder, McKenzie Wark and Faith Wilding (eds) (1999) *README!: ASCII Culture and the Revenge of Knowledge*, Autonomedia, New York. An interesting analysis of nettime is the essay 'Observations on Collective Cultural Action' (Critical Art Ensemble (2001) *Digital Resistance: Explorations in Tactical Media*, Autonomedia, New York, p. 71): 'Nettime', they write, 'functions as an information gift economy. Articles and information are distributed free of charge to members by those who have accumulated large information assets. Nettimers often see significant works on the intersections of art, politics and technology long before those works appear in the publications based on money economy.' In the spirit of this info gift economy, the whole book can be downloaded for free from http://www.critical-art.net.

12 B92 is at http://www.b92.net.

13 An archive of the Help B92 campaign is at http://www.helpb92.xs4all.nl.

14 One of a number of interviews archived at http://www.helpb92.xs4all.nl/indexe.html.

15 Collin, *This Is Serbia Calling*, p. 115.

16 http://www.freeb92.net/netaid/5/netaid.html.

17 For an account of B92's eventual liberation and the overthrow of Milosevic, see Collin, *This Is Serbia Calling*, chapter 7.

18 Constance Penley and Andrew Ross (1991) offer an interesting discussion on this point in their introduction to the edited collection *Technoculture*, University of Minnesota Press, Minneapolis and Oxford, pp. viii–xvii.

19 McKenzie Wark (1994) *Virtual Geography*, Indiana University Press, Bloomington, pp. 105–108.

20 Ali Mohammadi (1995) 'Cultural Imperialism and Cultural Identity', in Downing, Mohammadi and Sreberny-Mohammadi (eds) *Questioning The Media*, pp. 362–78.

21 Julianne Schultz (1994) 'Universal Suffrage? Technology and Democracy', in Lelia Green and Roger Guinery (eds) *Framing Technology*, Allen & Unwin, St Leonards, NSW, pp. 105–16.

22 Dorothy Denning (1999) 'Activism, Hacktivism and Cyberterrorism: the Internet as a Tool for Influencing Foreign Policy', http://www.nautilus.org/info-policy/workshop/papers/denning.html.
23 See 'How the Internet Helped Activists' (1998) *Straits Times*, 25 May, p. 14.
24 See John Larkin (1997) 'Student Radicals Regroup on Net', *Sydney Morning Herald*, 5 September, p. 14.
25 *VIP Reference* claims to have reached a quarter of a million people in this way. It is also sent to random addresses, so its recipients can argue to the authorities that they hadn't actually subscribed. See Denning, 'Activism, Hacktivism and Cyberterrorism'.
26 Collin, *This Is Serbia Calling*, p. 114.
27 John Keane (1991) *The Media and Democracy*, Polity Press, Cambridge, UK.
28 Keane, *The Media and Democracy*, p. 90.
29 Keane, *The Media and Democracy*, p. 91. See also Barbara Trent, 'Media in a Capitalist Culture'.
30 On the Pacifica crisis, see the postscript to Matthew Lasar (2000) *Pacifica Radio: the Rise of an Alternative Network* (updated edition), Temple University Press, Philadelphia, and John Downing et al (2001) *Radical Media: Rebellious Communication and Social Movements*, chapter 21. See also Barbara Epstein (1999) 'Radical Radio Fights to be Heard', *Le Monde Diplomatique*, October. The Free Pacifica Campaign http://www.radio4all.org/freepacifica archives a lot of information about the issues, as does the Save Pacifica site: see, for example, Matthew Lasar (1999) 'The Story So Far', at http://www.savepacifica.net/index.htm. Streaming audio of KPFA broadcasts is available at http://www.kpfa.org.
31 Lasar, *Pacifica Radio*, p. 44.
32 Evelyn Nieves (1999) 'Ever a Voice of Protest, Radio KPFA Is at It Again, but With a Twist', *New York Times*, 30 June, p. A12.
33 Lasar, *Pacifica Radio*, p. 181.
34 Lasar, *Pacifica Radio*, p. 220.
35 Although this has often proved difficult to put into practice. See Lasar, *Pacifica Radio*, chapter 8. These aspects of the structure of KPFA are also examined in Downing, *Radical Media*, chapter 21.
36 Lasar, *Pacifica Radio*, pp. 231–52.
37 Edward Herman (1999) 'Pacifica: Notes on Managerial Dysfunctionality', http://www.radio4all.org/fp/0803edherman.htm.
38 B92 sent KPFA a message of solidarity during the crisis, which was widely circulated online. The parallels between the two crises were drawn in a support site created by cultural saboteurs ®™ark, whom we'll meet properly in chapter 5. This site, which copies the design of the Help B92 site to emphasise the parallels, is at http://helpkpfa.rtmark.com.
39 Free Pacifica, http://www.radio4all.org/fp. It should be noted that this is only one of several factions created by the Pacifica crisis. Also of interest, for instance, is Save Pacifica, at http://www.savepacifica.net/index.htm.
40 This wire service, the A-Infos Radio Project, is at http://www.radio4all.net.
41 McSpotlight is at http://www.mcspotlight.org.
42 http://www.mcspotlight.org/campaigns/current/mcspotlight/faq.html.

43 Quoted in a July 1997 interview with the makers of the documentary *McLibel: Two Worlds Collide*, directed by Franny Armstrong and featuring courtroom reconstructions by Ken Loach. The video is available from http://www.spanner.org/mclibel. That interview with Dave Morris is at http://www.mcspotlight.org/people/interviews/morris97.html.

44 Dave Morris says, for example, that, according to McDonald's US Annual Report, the corporation's annual advertising budget in 1995 was $1.8 billion.

45 See http://www.mcspotlight.org/company/other_mclibels/index.html. The entire text of the play is online at McSpotlight, along with all legal correspondence between McDonald's and the theatre group.

46 London Greenpeace, founded in 1971, has no connection with its better-known namesake.

47 The leaflet is at http://www.mcspotlight.org/case/pretrial/factsheet.html.

48 The site is an ongoing project, entirely produced by collaboration between volunteers who refer to themselves as the McInformation Network. Like B92, these volunteers were helped to set up their site by XS4ALL – this meant that, from the beginning, McSpotlight was hosted in a different jurisdiction from that which applied to the McLibel defendants.

49 John Vidal (1997) *McLibel: Burger Culture on Trial*, Macmillan, London.

50 See, for example, David Shenk (1997) *Data Smog*, Abacus, London.

51 Tim Jordan (1999) *Cyberpower: the Culture and Politics of Cyberspace and the Internet*, Routledge, London, p. 101.

52 Jordan, *Cyberpower*, pp. 117–27.

53 An archive of the Blue Mountains campaign is at http://www.mcspotlight.org/campaigns/current/residents/blue_mountains.html.

54 Vidal, *McLibel*, p. 337.

55 In the US, for instance, the Census Bureau made a street-level database named Tiger available in 1994; this was used in producing mapping websites. See Walter R. Baranger (1997) 'Mapping Real World Gets Easier on the Web', *New York Times*, 26 May, p. 34.

56 Hybrid Media Lounge is at http://www.medialounge.net. While the project is ongoing, the data presented is a snapshot of February 1999.

57 Factory Watch is at http://www.foe.co.uk/campaigns/industry_and_pollution/factorywatch.

58 See Downing, 'Computers for Political Change: Peacenet and Public Data Access'.

59 See Peter Kollock (1999) 'The Economies of Online Cooperation', in Marc A. Smith and Peter Kollock (eds) *Communities in Cyberspace*, Routledge, London and New York, pp. 220–39.

60 http://www.mcspotlight.org/case/trial/verdict/verdict10_sum.html.

61 http://www.mcspotlight.org/media/press/msc_31mar99.html

62 This now-legendary remark is generally attributed to Electronic Frontier Foundation co-founder John Gilmore. For his views on this, see his site at http://www.toad.com/gnu.

63 http://www.mcspotlight.org/people/interviews/morris97.html.

64 A forum on corporate counter-activism, discussing the example of Shell, was held at the *Next 5 Minutes* tactical media conference in January 1999 (archived

at http://www.n5m.org/n5m3/pages/programme/counter.html). See also the Public Relations Society of America forum on countering anti-corporate activism, made available through the ®™ark website: http://www.rtmark.com/prsa. For a fuller discussion of corporate counter-activism in relation to online environmental campaigns, see Jenny Pickerill (forthcoming) *Cyberprotest: Environmental Activism Online*, Manchester University Press, Manchester, UK.

65 See http://www.prwatch.org.

66 Quote from the now-defunct website of the National Smokers Alliance, http://www.speakup.org, accessed 10 January 2000.

Chapter 4

1 Figures from the official Olympics site http://www.olympics.com. Online, the official Games website was perhaps the most ambitious and elaborate Net venture yet attempted, using almost 13 million lines of software code. Capitalising on NBC's decision to show no live coverage in the US, organisers were to claim some 11.3 billion hits by the end of competition. IBM, stung by failures in Atlanta, had spared no expense – some races, for example, could be followed online in close to real time, tracking competitors through microchips built into their running shoes. See Kirsty Needham (2000) 'Games Sites Net 30 Million Visitors', *Sydney Morning Herald*, 3 October, p. 3; Stewart Taggart (2000) 'NBC Delay Means Surf's Up in Oz', *Wired News*, 27 September.

2 Sydney Independent Media Centre, http://www.sydney.indymedia.org.

3 This policy is slightly porous, however: of nearly 300 stories posted to the end of August 2000, for example, some five had been removed for being completely off-topic; one example was a story about allegedly corrupt judges in the US, which the organisers felt had no relevance in a forum for issues arising from the Sydney Olympics. But even these stories are not completely deleted – they are still available on the site in a separate section. For a discussion of open publishing, see Matthew Arnison (2001) 'Open Publishing is the Same as Free Software', http://www.cat.org.au/maffew/cat/openpub.html.

4 See Matthew Arnison (2001) 'Crazy Ideas for Webcasting', http://www.cat.org.au/cat/webcast.html.

5 Alison Cordero (1991) 'Computers and Community Organizing: Issues and Examples from New York City', in John Downing, Rob Fasano, Patricia A. Friedland, Michael F. McCullough, Terry Mizrahi and Jeremy J. Shapiro (eds) *Computers for Social Change and Community Organizing*, Haworth Press, New York and London, pp. 89–103.

6 See, for example, Sharon Docter and William H. Dutton (1998) 'The First Amendment Online: Santa Monica's Public Electronic Network', in Rosa Tsagarousianou, Damian Tambini and Cathy Bryan (eds) *Cyberdemocracy: Technology, Cities and Civic Networks*, Routledge, London and New York, pp. 125–51. On the Santa Monica example, see also Howard Rheingold (1994) *The Virtual Community*, Minerva, London, pp. 268–72.

7 Christopher Mele (1999) 'Cyberspace and Disadvantaged Communities: the Internet as a Tool for Collective Action', in Marc A. Smith and Peter Kollock (eds) *Communities in Cyberspace*, Routledge, New York and London, pp. 290–310.

8 Sean Dodson and Patrick Barkham (2000) 'Protesters Limber up for Olympics', *The Guardian*, 19 July.

9 In making this point, I'm not trying to claim Sydney as the 'real centre' of the IMC movement, but rather to show that the movement has no single core. IMC programmer Matthew Arnison notes that while the majority of the IMCs use the same software package, one or two have experimented with variations: the Philadelphia IMC adapted code from Slashdot, while Quebec originally used commercial software, before moving towards the common IMC model. For May Day 2000 in London, organisers first used a different system before switching to the IMC standard.

10 Active Sydney, http://www.active.org.au/sydney.

11 Originating in San Francisco in 1993, Critical Mass is not an organisation, but a regular 'organised coincidence' in which as many cyclists as possible ride together through peak-hour traffic, advocating a range of aims, including sustainability, clean air and car-free public spaces. As with Reclaim the Streets, the focus is less on public *transport* than on public *space*: see John Jordan (1998) 'The Art of Necessity: the Subversive Imagination of Anti-Road Protest and Reclaim the Streets', in George McKay (ed.) *DiY Culture: Party & Protest in Nineties Britain*, Verso, London, pp. 129–51. Critical Mass Sydney is at http://www.nccnsw.org.au/~cmass/index.shtml.

12 Community Activist Technology, http://www.cat.org.au.

13 Eric S. Raymond (1997) 'The Cathedral and the Bazaar', http://www.tuxedo.org/~esr/writings/cathedral-bazaar.

14 A map of its development, 'The Roots and Shoots of the Active Software', is at http://www.active.org.au/doc/roots.pdf.

15 Email lists had been established before the website was launched; the first event on the web calendar was Survival Day, 26 January 1999.

16 Full list at http://www.sydney.active.org.au/groups.

17 A fascinating participant account of one project which successfully grappled with those difficulties is Martin Lucas and Martha Wallner (1993) 'Resistance by Satellite: the Gulf Crisis Project and the Deep Dish Satellite TV Network', in Tony Dowmunt (ed.) *Channels of Resistance: Global Television and Local Empowerment*, British Film Institute in association with Channel Four Television, London, pp. 176–94.

For a different take on this kind of potential, see Nicholas Garnham (1990) 'The Myths of Video: a Disciplinary Reminder', in his *Capitalism and Communication*, Sage, London, pp. 64–69. In a forceful critique of the empowering potential of new media – in this case, video cameras – Garnham ridicules the idea that access to video technology would also allow its users access to the established TV networks. But he ignores the more interesting – and more real – possibility of the establishment of *alternative networks*. While *Undercurrents* had put this potential into action, the IMC network takes it to another level: the footage can be viewed from any suitably equipped entry-level computer (although whether a collective public screening/meeting is better in terms of stimulating debate is an arguable point). An irony of Garnham's critique is that in emphasising the role of marketing hype and powerful economic interests in spreading his 'myths of video', he denies the autonomy of those who are able to see past this hype.

18 On the political resonances of major sporting events see, for example, Jock Given (1995) 'Red, Black, Gold to Australia: Cathy Freeman and the Flags', *Media Information Australia*, no. 75, pp. 46–56.

19 Jon Savage (1991) *England's Dreaming: Sex Pistols and Punk Rock*, Faber & Faber, London, p. 401.

20 Ibid., p. 202.

21 George McKay (1998) 'DiY Culture: Notes Towards an Intro', in George McKay (ed.) *DiY Culture*, pp. 1–53, quote from p. 2.

22 Stephen Duncombe (1997) *Notes From Underground: Zines and the Politics of Alternative Culture*, Verso, London, p. 190.

23 Annabel McGilvray (2000) 'Protesters Rally Games Sit-In on Net', *The Australian*, 7 August, p. 4.

24 Andrew Clennell (2000) 'Anarchist Protest Plan Revealed', *Sydney Morning Herald*, 7 August, p. 4.

25 Todd Gitlin (1980) *The Whole World Is Watching: Mass Media in the Making and Unmaking of the New Left*, University of California Press, Berkeley, pp. 122–23. While Gitlin's comments pre-date the trend towards separate commentary and analysis supplements in newspapers, they still apply to the main body of news reports.

26 Elizabeth S. Bird and Robert W. Dardenne (1997) 'Myth, Chronicle and Story: Exploring the Narrative Qualities of News', in Dan Berkowitz (ed.) *Social Meanings of News*, Sage, Thousand Oaks, pp. 333–50.

27 Stuart Hall (1984) 'The Narrative Construction of Reality: an Interview', *Southern Review*, vol. 17, no. 1, pp. 3–17, quote from p. 5.

28 Charles R. Bantz (1997) 'News Organizations: Conflict as a Crafted Cultural Norm', in Berkowitz (ed.) *Social Meanings of News*, pp. 123–37, quote from p. 133.

29 Bird and Dardenne, 'Myth, Chronicle and Story', p. 333.

30 Hall, 'The Narrative Construction of Reality', p. 7.

31 Robert Stam (1983) 'Television News and Its Spectator', in E. Ann Kaplan (ed.) *Regarding Television*, University Publications of America, Frederick, MD, pp. 23–43, quote from p. 31.

32 For a discussion of the need for new types of journalistic and political narratives, see W. Lance Bennett and Murray Edelman (1985) 'Toward a New Political Narrative', *Journal of Communication*, vol. 35, no. 4, pp. 156–71. Bennett and Edelman propose an approach 'designed to focus contradictions and normative dilemmas within the same story' (p. 170). Achieving this would involve recognising 'the inevitability of contradictory stories, the multiple realities they evoke, and their links to the conditions of people's lives' (p. 171).

33 'e-Life' (cover story) (1999) *Newsweek*, 11 October.

34 On this question see (inter alia) Langdon Winner (1977) *Autonomous Technology*, MIT Press, Cambridge, MA.

35 Merritt Roe Smith and Leo Marx (1994) 'Introduction', in Merritt Roe Smith and Leo Marx (eds) *Does Technology Drive History? The Dilemma of Technological Determinism*, MIT Press, Cambridge, MA, p. x.
36 Langdon Winner (1999) 'Who Will We Be In Cyberspace?', in Paul A. Mayer (ed.) *Computer Media and Communication: a Reader*, Oxford University Press, pp. 207–18, quote from pp. 216–17.
37 Nelly Oudshoorn (1996) 'A Natural Order of Things?', in George Robertson, Melinda Mash, Lisa Tickner, Jon Bird, Barry Curtis and Tim Putnam (eds) *Future Natural: Nature, Science, Culture*, Routledge, London and New York, pp. 122–32.
38 An excellent account of Telefon Hirmondó is found in Carolyn Marvin (1988) *When Old Technologies Were New: Thinking About Communications in the Late Nineteenth Century*, Oxford University Press (OUP), Oxford.
39 John Naughton (1999) *A Brief History of the Future*, Weidenfeld & Nicolson, London, p. 245.
40 Richard Dyer (1997) *White*, Routledge, London and New York, p. 89.
41 Ibid., p. 90.
42 See the interview with Eco in Lee Marshall (1997) 'The World According to Eco', *Wired* 5.03, March, http://www.wired.com/wired/archive/5.03/ff_eco.html.
43 Steven Levy (1994) *Insanely Great*, Penguin, Harmondsworth, pp. 68–70.
44 See Geert Lovink (2000) 'Cyberculture in the Age of Dotcom Mania', posted to the nettime list 15 April. Archived at http://www.nettime.org.
45 William Gibson (1986) 'Burning Chrome', collected in (1995) *Burning Chrome and Other Stories*, HarperCollins, London, p. 215.

Besides the issue of user adaptation, it's also important to note that such in-built politics don't always go unchallenged – users and publics can take an active role. Monsanto, for example, created a huge outcry with their development of products using the so-called Terminator gene – self-sterilising seeds which could be used only once, forcing farmers into an annual commercial dependence on their supplier (such commercial emphases, we should note, are central to biotechnology: as one science journalist notes, the very word 'biotechnology' was invented on Wall Street. See Robert Teitelman (1989) *Gene Dreams: Wall Street, Academia and the Rise of Biotechnology*, Basic Books, New York, p. 4). Along with the image of 'Frankenstein foods', the Terminator became a metonym for genetic modification, a focal point for opposition. In India, farmers torched fields planted with genetically modified (GM) crops as part of 'Cremate Monsanto Day'. In parts of Japan, campaigners succeeded in having GM foods banned from school lunches. In the UK, all major supermarket chains removed GM products from their own-brand food ranges, with similar action taken in other EU countries, including Germany, Italy and France. Clearly, this is not a technology whose implementation is proceeding as planned. The response to these particular applications of biotechnology is a striking instance of a grassroots rejection of developments whose benefits were apparent to nobody but the stockholders of the biotech corporations involved. Sources on international reactions: Katharine Inez Ainger (1999) 'The Meek Fight For Their Inheritance', *Guardian Weekly*, 21 February, p. 23; Sonni Efron

(1999) 'Food Activists Fight For Raw Deal', *Sydney Morning Herald*, 20 March, p. 23; John Vidal (1999) 'We're Gagging on GM', *The Guardian*, 19 March.

46 On hip-hop and the turntable, see Craig Werner (1998) *A Change Is Gonna Come: Music, Race & the Soul of America*, Payback Press, Edinburgh, chapter 43; and Houston A. Baker, Jr. (1991) 'Hybridity, the Rap Race, and Pedagogy for the 1990s', in Constance Penley and Andrew Ross (eds) *Technoculture*, University of Minnesota Press, Minneapolis, pp. 197–209. On house music and the Roland 303, see Simon Reynolds (1998) *Energy Flash: a Journey Through Rave Music and Dance Culture*, Picador, London, chapter 1.

47 Langdon Winner (1986) *The Whale and the Reactor*, University of Chicago Press, Chicago and London, pp. 9–10.

48 Franklin, quoted in Merritt Roe Smith (1994) 'Technological Determinism in American Culture', in Smith and Marx (eds) *Does Technology Drive History?*, pp. 1–35, quote from p. 3.

49 On open source software see http://www.opensource.org, http://www.gnu.org and http://www.slashdot.org. For an accessible history of the movement see John Naughton (1999) *A Brief History of the Future*, Weidenfeld & Nicolson, London. See also Lawrence Lessig (1999) *Code*, Basic Books, New York.

50 See, for example, Glyn Moody (2001) *Rebel Code*, Penguin, Harmondsworth. The open source/hacker ezine Slashdot is an important precursor to the IMC model, although its commitment to open publishing is less rigorous: http://www.slashdot.org. For another precursor to the IMC open publishing model, see Atton's discussion of the experimental publishing policy of *Green Anarchist* magazine: Chris Atton (2002) *Alternative Media*, Sage, London, chapter 5.

51 Graham Lawton (2002) 'The Great Giveaway', *New Scientist*, 2 February, is also at http://www.newscientist.com/hottopics/copyleft/copyleftart.jsp.

52 Tomoko Takahashi and Jon Pollard (2000) *Word Perhect* http://www.frieze.com/projects/perhect/frame.html.

53 Jeanette Hofmann (1999) 'Writers, Texts and Writing Acts: Gendered User Images in Word Processing Software', in Donald MacKenzie and Judy Wajcman (eds) *The Social Shaping of Technology* (2nd edition), Open University Press, Buckingham and Philadelphia, pp. 222–43. On the politics of word processing, see also Matthew Fuller (2000) 'It Looks Like You're Writing a Letter: Microsoft Word', at http://www.axia.demon.co.uk/wordtext.html.

54 Natural Selection http://www.mongrelx.org/Project/Natural. See also Matthew Fuller's accompanying essay, 'The War of Classification', at http://www.axia.demon.co.uk/woc.html.

55 Naughton, *A Brief History of the Future*, pp. 32–33. The *Time* magazine cover story was: Philip Elmer-DeWitt (1995), 'On a Screen Near You: Cyberporn', 3 July, archived at http://www.time.com/time/magazine/archive. The Electronic Frontier Foundation website has a good archive of material on the case at http://www.eff.org/Censorship/Rimm_CMU_Time.

56 Stephen Duncombe (1997) *Notes From Underground: Zines and the Politics of Alternative Culture*, Verso, London, p. 142.

57 Craig Mathieson (2000) *The Sell-In: How the Music Business Seduced Alternative Rock*, Allen & Unwin, St Leonards, NSW.

58 Naomi Klein (2000) *No Logo*, Flamingo, London. On this point see particularly chapter 5, 'Patriarchy Gets Funky: the Triumph of Identity Marketing'.

Chapter 5

1 ®™ark are at http://rtmark.com. At the time of writing, in March 2002, Exley continues to run his Bush site, although ®™ark are no longer directly involved: http://www.gwbush.com. A complete archive of the ®™ark content for gwbush.com, including the huge amount of press coverage and the legal correspondence from Bush's attorneys, is at http://rtmark.com/bush.html.

2 A similar program – Reamweaver – subsequently became available for anyone to download from the site of the Yes Men, http://www.theyesmen.org.

3 For Phone In Sick Day, an entertaining archive of press coverage ('Most Irish Police Off Sick', etc.) is at http://rtmark.com/presspis.html. The annual Corporate Poetry prize is awarded to real examples of corporate globospeak. Particularly recommended is the 1998 winner, from the Brookings Institution, that begins 'I like markets . . .', http://rtmark.com/corpoetry.html.

4 A helpful guide to DIY Barbie surgery appears in Jennifer Terry and Melodie Calvert (eds) (1997) *Processed Lives: Gender and Technology in Everyday Life*, Routledge, London and New York, pp. 196–97. You'll need a soldering iron.

Instructions on similar, if more technically ambitious, projects – reverse engineering the Nintendo GameBoy and constructing robotic graffiti-writers – are detailed in Critical Art Ensemble (2001) *Digital Resistance: Explorations in Tactical Media*, Autonomedia, New York. Other 'contestational robotics' can be found at the Institute for Applied Autonomy, at http://www.appliedautonomy.com.

5 Quote from an ®™ark press release at http://rtmark.com/bushpr2.html.

6 A tiny precedent might be the UK's Tom Bisson, who, in what was no doubt a very moving ceremony, married his local pub: Kendall Hill (2001) 'Modern History', *Sydney Morning Herald*, 3 March, p. 30. See also the exchange of love letters between Daniel Arp and Amazon.com's customer service department (a co-winner of ®™ark's 2000 Corporate Poetry Contest) at http://www.deadletters.com/inbox/amazon.html.

7 The parody site, originally at http://www.realjeff.com, is still available online at the National Library of Australia's Pandora project, which archives significant Australian websites; Kennett's official site is not included. Pandora is at http://pandora.nla.gov.au/index.html.

We should note that such clone sites are by no means the exclusive domain of the Left. See, for instance, the site titled 'Martin Luther King, Jr: An Historical Examination' at http://www.martinlutherking.org. The site presents itself at first glance as pro-King, featuring headings such as 'Civil Rights Library' and 'The King Holiday: Bring the Dream to Life'. But it is in fact an extended character assassination and historical revisionism project, with links to former Ku Klux Klan director David Duke.

8 As the campaign progressed, Exley broadened the emphasis from the cocaine issue to capital punishment, pointing to what he saw as the inconsistency between Bush's slogan of 'compassionate conservatism' and the packed execution schedule in his home state of Texas during election year.

The term 'brochureware' describes a website which does no more than duplicate the kinds of material available elsewhere (such as in a print brochure) without taking advantage of the properties of the web.

9 GATT was the previous name for the WTO.

10 Moore, quoted in http://www.wto.org/english/news_e/pres99_e/pr151_e.htm. The illegality of such parody sites is a grey area, but probably won't be for long. See, for instance, this proposed legal model, which singles out several ®™ark projects as examples of – I kid you not – 'commercial terrorism'. Bruce Braun, Dane Drobny and Douglas C. Gessner (2000) 'www.commercial_terrorism.com: a Proposed Federal Criminal Statute Addressing the Solicitation of Commercial Terrorism Through the Internet', *Harvard Journal on Legislation*, vol. 37, no. 1, pp. 159–85.

11 Ray Thomas in http://rtmark.com/gattpr.html. See, again, the Yes Men site for more on subsequent developments in this particular example.

12 See David Garcia and Geert Lovink (1997) 'The ABC of Tactical Media', http://www.waag.org/tmn/frabc.html. It's worth noting that not all tactical media practitioners are keen on the label – in their latest book, *Digital Resistance: Explorations in Tactical Media* (2001, Autonomedia, New York), Critical Art Ensemble point out that naming and defining tactical practices leaves them open to being co-opted: 'Should tactical media become popularized, its recuperation by capital is almost inevitable' (p. 5). I think this is true, and acknowledge that my own discussion of tactical media in this chapter is part of the process of academic recuperation. But in using the phrase 'tactical media' in the subtitle of their book, Critical Art Ensemble are also contributing to this process, as they are no doubt aware.

13 See the FAQ section at http://www.n5m.org.

14 Geert Lovink and Florian Schneider (2001) 'New Rules of the New Actonomy', http://new.actonomy.org.

15 Critical Art Ensemble, *Digital Resistance*, p. 63.

16 Michel de Certeau (1984) *The Practice of Everyday Life*, University of California Press, Berkeley, Los Angeles and London. Translated by Steven F. Rendall.

17 Ibid., p. 31. Compare with Umberto Eco's essay on 'semiological guerilla warfare': 'The battle for the survival of man as a responsible being in the Communications Era is not to be won where the communication originates, but where it arrives'; Umberto Eco (1987) [1967] 'Towards a Semiological Guerilla Warfare', in *Travels in Hyperreality*, Picador, London, pp. 135–44, quote from p. 142.

On the question of what audiences *do* with messages, see Marie Gillespie (1995) *Television, Ethnicity and Cultural Change*, Routledge, London and New York. Gillespie examines the roles played by soap operas in the lives of British teenagers of Punjabi background. While their parents worry that their kids may mindlessly copy what they see on the screen and hence become 'Western', the kids themselves describe how gossip about soap storylines helps them work out some of their own problems without having to voice them. The soap characters offer proxies, and gossiping about the characters' problems with school, work, parents and relationships gives the teenagers a vehicle to explore and form their own opinions about issues in their own lives. The students Gillespie interviews don't *identify* with the soap characters so much as they *associate* the characters with their own lives – they make the shows similar to their lives, rather than becoming similar to the shows. Making sense of what they've watched is less a process of imbibing some set reading, and more a process of, in Gillespie's words, 'negotiation and struggle' (p. 149). See also Ien Ang (1985) *Watching Dallas*, Methuen, London, and (1991) *Desperately Seeking the Audience*, Routledge, London; and John Fiske (1989) *Television Culture*, Routledge, London.

18 De Certeau, *The Practice of Everyday Life*, p. xix.

19 Ibid., p. 37.

20 Hakim Bey (1991) *T.A.Z.: the Temporary Autonomous Zone, Ontological Anarchy, Poetic Terrorism*, Autonomedia, New York, p. 100.

21 A good introduction to the enormous range of self-produced zines made possible by photocopiers is V. Vale (1996) *Zines!* Volume 1 and (1997) Volume 2, both V/Search, San Francisco. See also Stephen Duncombe (1997) *Notes From Underground: Zines and the Politics of Alternative Culture*, Verso, London and New York, and Chris Atton (2002) *Alternative Media*, Sage, London, chapter 3. Zine directories online often go without updates, but some starting points, some of which aren't being maintained any more but still have archival value, include *Factsheet 5* http://www.factsheet5.com; *The Zines, E-Zines Resource Guide* http://www.zinebook.com; *The New Pollution* http://evolver.loud.org.au/nupoo/about.

22 See, for example, Greg Ruggiero (1999) *Microradio & Democracy: (Low) Power to the People*, Seven Stories Press, New York. On pirate radio, see Simon Reynolds (1998) *Energy Flash*, Picador, London, chapter 9, 'This Sound Is For The Underground', and Steve Lowe (2001) 'Pirate Radio Ahoy!', *Q*, no. 183, pp. 86–92.

23 See, for example, Philip Cornford (1999) 'Global Mobile Terrorists', *Sydney Morning Herald*, 18 February, p. 1.

24 An interview with the artist responsible, Krysztof Wodiczko, and some good pictures of his work, is in Louisa Buck (1988) 'Bright Lights, Big Cities', *The Face*, no. 98, pp. 48–51.

25 The imagery was also later projected at other sites in Sydney. See http://www.boat-people.org.

26 The messages to be projected were submitted, selected and co-ordinated through the now-defunct website Hello Mr President, formerly at http://hellomrpresident.com, accessed 18 December 2001.

27 This event, known as both National Accountability Day and ShootBack Day, tries to draw attention to the increasing surveillance of public space: see http://wearcam.org/nad-faq.htm.

28 When and if they have five million registered citizens, Cyber Yugoslavia intend to apply for membership of the UN. On acceptance, they'll apply for 20 square

metres of territory somewhere in the world on which to place their web server. See Cyber Yugoslavia, http://www.juga.com.

29 *Undercurrents* is an independent video news magazine, found at http://www.undercurrents.org. For press coverage of the Shell action see John Vidal (1999) 'Warfare Across the Web', *Guardian Weekly*, 7 February, p. 20. An insider's account of the development of *Undercurrents* is Thomas Harding (1998) '*Viva Camcordistas!* Video Activism and the Protest Movement', in George McKay (ed.) *DiY Culture: Party & Protest in Nineties Britain*, Verso, London, pp. 79–99.
30 A fascinating time capsule of the Greenham occupations is Alice Cook and Gwyn Kirk (1983) *Greenham Women Everywhere*, Pluto, London.
31 See, for example, the website of the Electronic Disturbance Theater (whom we'll meet properly in chapter 6), http://www.thing.net/~diane/ecd/ztps.html. See also John Jordan and Jennifer Whitney (2001) 'Resistance is the Secret of Joy', *New Internationalist*, no. 338, http://www.oneworld.org/ni/issue338/secret.htm. Earlier examples of such 'nonviolent air raids' can be found in Gene Sharp (1973) *The Politics of Nonviolent Action* (in 3 volumes), Porter Sargent, Boston, pp. 381–82.
32 De Certeau, *The Practice of Everyday Life*, p. 37.
33 John Ralston Saul (1995) *The Doubter's Companion*, Penguin, Harmondsworth, p. 67.
34 Ibid., p. 65.
35 Saul D. Alinsky (1971) *Rules For Radicals: a Pragmatic Primer for Realistic Radicals*, Vintage Books, New York, p. 75.
36 Ibid., pp. 137–38.
37 Ibid., p. 128.
38 Mustapha Khayati (1981) [1966] 'Captive Words: Preface to a Situationist Dictionary', in Ken Knabb (ed.) *Situationist International Anthology*, Bureau of Public Secrets, Berkeley, pp. 170–75, quote from p. 170. The full text of Knabb's anthology is also online at http://www.bopsecrets.org/SI/contents.htm.
39 Mark Dery (1993) 'Culture Jamming: Hacking, Slashing and Sniping in the Empire of Signs', full text online at http://www.levity.com/markdery/culturjam.html. Not all culture jamming, we should note, lives up to this ideal.
40 Available online in streaming video, this film – *Bringing It To You!* – is a funny and deadly accurate parody of corporate self-promotion. It's found at http://rtmark.com/vidbity.html.
41 Arjun Appadurai (1996) *Modernity At Large: Cultural Dimensions of Globalization*, University of Minnesota Press, Minneapolis, pp. 27–47.
42 Barme and Jaivin, quoted in McKenzie Wark (1994) *Virtual Geography: Living With Global Media Events*, Indiana University Press, Bloomington, p. 97. On the historical shifts and multiple meanings of terms like 'reform' and 'democracy' see Raymond Williams (1976) *Keywords*, Fontana, London.
43 For example, in the ouster of Indonesian President Soeharto in 1998, the key word was 'reform'. Everyone used it, but no one seemed to agree on what it meant. In early May students in Jakarta called for 'total reform'. A politics professor sympathetic to the students' cause expressed his exasperation with

the vagueness of this demand: 'What they say is merely rhetorical. They never discuss details.' An army officer was quoted as saying 'We cannot tolerate total reform. We prefer gradual reform.' One *Jakarta Post* columnist commented that the officer 'did not explain what he meant by total and gradual', while an editorial on the same day observed that '"reform" as it is understood by those in power constitutes something different and far short from what government critics have in mind'. The decisive amplification of the term came after the fatal shootings of students at Trisakti University: the dead became 'martyrs in a holy war', later becoming 'Champions of Reform, and Flowers of Reform', while Trisakti itself was now 'the Campus of Reform'. Spotting the trend, but clearly failing to grasp it, Soeharto offered to create a 'reform cabinet', to be appointed by himself and to include his daughter. By late May the *Jakarta Post* could observe that the word had become 'something of a national motto', while an Australian correspondent made the valuable point that 'The poor in the villages think of *reformasi* not in terms of democracy and human rights, but in the cost of cooking oil and rice, and the value of their crops.' Sources: 'Students Extend Demand on Total Govt Reforms' (1998) *Jakarta Post*, 2 May, p. 2; 'Is There Any End to Student Demos?' (1998) *Jakarta Post*, 6 May, p. 1; Marsilam Simanjuntak (1998) 'Transitional Leadership Is Needed For Reform', *Jakarta Post*, 13 May, p. 1; 'Expediting Reform' (1998) *Jakarta Post*, 13 May, p. 4; 'Students Honor "Reform Heroes"' (1998) *Jakarta Post*, 14 May, p. 1; Ivy Susanti (1998) 'Students Remembered as Reform Marches Forward', *Jakarta Post*, 27 May, p. 3; Louise Williams (1998) 'Collision Course', *Sydney Morning Herald*, 20 May, p. 1; 'Reforming Attitudes' (1998) *Jakarta Post*, 22 May, p. 4; Greg Bearup (1998) 'Joy and Fear in the Paddy Field', *Sydney Morning Herald*, 22 May, p. 8.

44 This slide show – another past winner of ®™ark's Corporate Poetry contest – is at http://www.monsanto.com.

45 Adbusters, http://www.adbusters.org.

46 Adbusters' spoof ads gallery is at http://adbusters.org/spoofads. A site devoted to collecting and presenting examples of subvertisements is http://www.subvertise.org.

47 http://www.adbusters.org/campaigns/bnd/toolbox/update00.

48 http://www.adbusters.org/campaigns/bnd.

49 Craig Baldwin (1995) *Sonic Outlaws*. Itself an illustration of culture jammers' aesthetics, Baldwin's film examines the legalities of sampling and copyright through an account of a dispute between Negativland and U2's record company. Another participant in Baldwin's film offers a definition of jamming which is identical to detournement: 'Jamming always supersedes found content as it rearranges it, causing you to reflect on the nature of what you're actually encountering.' Similarly, Frank Guerrero of ®™ark describes culture jamming as 'interrupting/amending/disrupting the established flow of information... snatching things and ideas from one context and placing them in another in order to reveal something about those ideas or contexts'.

50 Dery, 'Culture Jamming'. Still another way of defining culture jamming would be to take Greil Marcus's description of Dada as 'a marriage of prank and negation', and add to that a specific emphasis on opposition to corporate media

and an explicit objective of promoting media literacy. See Greil Marcus (1989) *Lipstick Traces: a Secret History of the Twentieth Century*, Picador, London, p. 33. One example from this perspective might be the mini-moral panic generated by the hoax social movement Young People Against Heavy Metal T-Shirts; its creator recounts his tale of credulous media in Matthew Thompson (1997) 'Tabloid Whore', *Australian Style*, January, pp. 78–80. Another might be the 'Dole Army' scam: on 4 February 2002, two Australian commercial TV current affairs shows broadcast interviews with members of the Dole Army, who claimed to be a 70-strong group that lived underground and scavenged for food while plotting new ways to extort welfare benefits; this troglodyte cabal was, of course, a hoax aimed at the preconceptions of journalists, who proved only too eager to believe. Video of an ABC *Lateline* report on the Dole Army hoax is archived at http://www.abc.net.au/lateline/av/2002/02/20020205llhoax.ram. See also V. Vale and Andrea Juno (1987) *RE/Search #11: Pranks!*, V/Search, San Francisco; and the website of veteran prankster Joey Skaggs, at http://www.joeyskaggs.com.

51 Dery, 'Culture Jamming'.

52 Critical Art Ensemble (1994) *The Electronic Disturbance*, Autonomedia, New York, p. 51.

53 This media literacy emphasis means that even some programs on network TV can also be seen as culture jamming: the current affairs satires of *Frontline* or *The Election Chaser*; the media pranks of Michael Moore's *The Awful Truth*; and Ali G's deadpan interviews with nonplussed experts on drugs, crime or animal rights. See also the two pilot episodes of John Safran's TV series which, though rejected by the ABC, remain available on the Net. *John Safran: Media Tycoon* and *John Safran: Master Chef* can be viewed at http://fuck-the-skull-of-jesus.mit.edu/safran.html. All of these both promote and rely on a critical understanding of, and engagement with, media conventions. In such current affairs satire, perhaps no one has gone further than evil genius Chris Morris in his UK series *Brass Eye*. In an episode on drugs he persuaded one irony-blind MP to ask a question in Parliament about the problem of 'cake', which Morris had told him was 'a made-up drug'. The MP seemed to think this meant 'designer drug'. But it meant made-up.

54 Guy Debord (1981) [1957] 'Report on the Construction of Situations and on the International Situationist Tendency's Conditions of Organization and Action', in Ken Knabb, *Situationist International Anthology*, pp. 17–25, quote from p. 17. A good online archive of Situationist writings is at http://www.nothingness.org.

55 Compare Buy Nothing Day, for example, with the December 1968 action at Selfridges in London, when 25 associates of the Situationist offshoot King Mob (including Malcolm McLaren) dressed up as Santa Claus and tried to give away the shop's toys to passing kids. See Jon Savage (1991) *England's Dreaming: Sex Pistols and Punk Rock*, Faber, London, p. 34.

56 Quoted in Knabb, *Situationist International Anthology*, p. 385.

57 Guy Debord (1987) [1967] *The Society of the Spectacle*, Rebel Press, Exeter, section 1.

58 Marcus, *Lipstick Traces*, p. 49.

59 Marcus, *Lipstick Traces*, p. 37.

60 *Detournement* is a word best left untranslated. Knabb argues that the precise meaning is not captured by any of the usual English equivalents. In her study of the Situationists, cultural studies academic Sadie Plant offers this definition: 'The closest English translation of *detournement* lies somewhere between "diversion" and "subversion". It is a turning around and a reclamation of lost meaning: a way of putting the stasis of the spectacle in motion. It is plagiaristic, because its materials are those which already appear within the spectacle, and subversive, since its tactics are those of the "reversal of perspective", a challenge to meaning aimed at the context in which it arises'; Sadie Plant (1992) *The Most Radical Gesture: the Situationist International in a Postmodern Age*, Routledge, London, p. 86.

61 Guy Debord and Gil Wolman (1981) [1956] 'Methods of Detournement', in Ken Knabb, *Situationist International Anthology*, pp. 8–14, quote from p. 9.

62 Umberto Eco (1987) [1978] 'Falsification and Consensus', in *Travels in Hyperreality*, pp. 173–79, quote from p. 174. But Eco is not a cheerleader for the prospects of this kind of activity. 'At most', he writes, 'it guarantees the mutual survival of the players of the game. The big publishing houses are ready to accept the spread of photocopying, as the multinationals can tolerate the phone calls made at their expense, and a good transportation system willingly accepts a fair number of counterfeit tickets – provided the counterfeiters are content with their immediate advantage' (p. 179).

63 Disinformation is at http://www.disinfo.com. An interview with Metzger and an interesting account of the site's early development is R.U. Sirius (1997) 'Media Pranksters', *21C*, no. 25, pp. 46–51. On Disinformation's mix of Left politics and conspiracy theory, we might note also Fredric Jameson's observation that conspiracy theory is one means of understanding the otherwise inexplicable nature of global capital in a systematic fashion: 'Conspiracy', he writes, 'is a degraded figure of the total logic of late capital, a desperate attempt to represent the latter's system'; Fredric Jameson (1988) 'Cognitive Mapping,' in Cary Nelson and Lawrence Grossberg (eds) *Marxism and the Interpretation of Culture*, University of Illinois Press, Urbana and Chicago, pp. 347–60, quote from p. 356.

64 *The Digital Landfill* http://www.potatoland.org/landfill.

65 *re-m@il* http://www.sero.org/cgi-bin/remail/re-mail.pl.

66 An interview with Mongrel and collaborators is Matthew Fuller (1999) 'The Mouths of the Thames', 14 February, archived at the nettime list, http://www.nettime.org.

67 Media Carta, http://www.adbusters.org/campaigns/mediacarta. The Universal Declaration of Human Rights is at http://www.un.org/Overview/rights.html.

68 A job which is done very well by alternative media website The Media Channel, http://www.mediachannel.org. These processes of corporate convergence are the subject of a number of major studies from within the political economy school of media studies. In their 1997 book *The Global Media*, for example, Herman and McChesney trace the ascendancies of the dominant media corporations, profiling the main global firms, sketching what they see as the

likely future absorption of the Internet into the corporate orbit, and attempting to assess the influence of globalising tendencies in the US and a number of other countries. The Media Carta campaign reflects the concerns raised by such critics. See Edward S. Herman and Robert W. McChesney (1997) *The Global Media: the New Missionaries of Corporate Capitalism*, Cassell, London and Washington. See also Ben H. Bagdikian (2000) *The Media Monopoly* (6th edition), Beacon Press, Boston; Edward S. Herman and Noam Chomsky (1988) *Manufacturing Consent: the Political Economy of the Mass Media*, Vintage, London.

69 On convergence, see Trevor Barr (2000) *newmedia.com.au*, Allen & Unwin, St Leonards, NSW.

70 Both quoted in Kalle Lasn (1999) *Culture Jam: the Uncooling of America™*, Eagle Brook, New York, pp. 32–33.

71 As you'll have noticed from my first mention of Adbusters, however, CNN is happy to take their money, and has aired the Buy Nothing Day ad since 1998, during a show which regularly reaches a million viewers. The ad also airs on community TV networks and college radio stations across the US.

72 Manuel Castells (1996) *The Rise of the Network Society*, Blackwell, Oxford, p. 35.

Chapter 6

1 See the contemporary report 'Protest: the Banners of Dissent' (1967) *Time*, vol. 90, no. 17, 27 October, pp. 9–15. An entertaining account of the levitation of the Pentagon is found in Larry Sloman (1998) *Steal This Dream: Abbie Hoffman and the Countercultural Revolution in America*, Doubleday, New York.

2 Woody Allen (1997) 'A Brief, Yet Helpful, Guide to Civil Disobedience', *Complete Prose*, Picador, London, p. 69.

3 I wrote these introductory paragraphs about the Pentagon on 10 September 2001. The next night I sat, like everyone else, in front of CNN as a jet crashed into the Pentagon building. I thought of changing that introduction – it suddenly seemed kind of distasteful to talk of 'provoking' the Pentagon. And perhaps it is. But I've kept it, because the events of 11 September do put hacktivism into some kind of perspective. Whatever hacktivism is – and, as we'll see, its meaning is still evolving – it's not terrorism. Just as 'hacking' now conjures up some stereotyped phantoms, the very concept of 'terrorism' actively obscures any understanding of the historical and contemporary context for acts of political violence. And, again as with 'hacking', media coverage of such acts frequently makes good on former UK Prime Minister John Major's extraordinary suggestion that 'Society needs to condemn a little more and understand a little less' (interviewed in the *Mail on Sunday*, 21 February 1993). On the difficulties of defining 'terrorism' see (inter alia) Conor Gearty (1997) *The Future Of Terrorism*, Phoenix, London, and Bruce Hoffman (1998) *Inside Terrorism*, Victor Gollancz, London. On the substantial differences between electronic civil disobedience and terrorism, see the Congressional Testimony on Cyberterrorism by Dorothy Denning (2000) at http://www.cs.georgetown.edu/~denning/infosec/cyberterror.html. See also 'The Mythology of Terrorism on the Net', in Critical

Art Ensemble (2001) *Digital Resistance: Explorations in Tactical Media*, Autonomedia, New York, pp. 29–37. The full text of this, and of their three other books, is available for free download at http://www.critical-art.net. For an example of the media drawing linkages between electronic civil disobedience and 'terrorism', see, from the *Washington Post* network, Kevin Featherly (2001) 'U.S. On Verge of "Electronic Martial Law" – Researcher', *Newsbytes*, 15 October, http://www.newsbytes.com/news/01/171130.html.

4 Mark Little (1999) 'Practical Anarchy: an Interview with Critical Art Ensemble', *Angelaki*, vol. 4, no. 2, pp. 192–201, quote from p. 194.

5 Distributed to a range of mailing lists, including nettime, where it's archived at http://www.nettime.org. The Electronic Disturbance Theater's site is at http://www.thing.net/~rdom/ecd/ecd.html.

6 See http://www.thing.net/~rdom/zapsTactical/warning/htm.

7 Manuel Castells (1997) *The Power of Identity*, Blackwell, Oxford, p. 79.

8 The origins and objectives of the Zapatista movement are too complex to go into here in any detail. An annotated directory of Zapatista Internet projects is academic Harry Cleaver's 'Zapatistas in Cyberspace: a Guide to Analysis & Resources', at http://www.eco.utexas.edu/faculty/Cleaver/zapsincyber.html. See also John Womack (1999) *Rebellion in Chiapas: an Historical Reader*, The New Press, New York. Two interesting analyses of the EZLN's communications strategies are in Manuel Castells, *The Power of Identity*, pp. 72–83, and John Arquilla and David Ronfeldt (1998) *The Zapatista Social Netwar in Mexico*, RAND Corporation report, available online at http://www.rand.org/piblications/MR/MR994.

9 Critical Art Ensemble (1995) *Electronic Civil Disobedience and Other Unpopular Ideas*, Autonomedia, New York, p. 15.

10 Critical Art Ensemble, *Electronic Civil Disobedience*, p. 18.

11 Bruce Sterling (1992) *The Hacker Crackdown*, Viking, London, p. 81.

12 Critical Art Ensemble (1994) *The Electronic Disturbance*, Autonomedia, New York, p. 26.

13 Steven Levy (1984) *Hackers: Heroes of the Computer Revolution*, Doubleday, New York, p. 208.

14 Sterling, *The Hacker Crackdown*, p. 230.

15 Henry David Thoreau (2000) [1849; 1866] 'Civil Disobedience', in Paul Lauter (ed.) *Walden and Civil Disobedience*, Houghton Mifflin, Boston, p. 18. Lauter's edition includes an excellent anthology of extracts from texts on civil disobedience by Tolstoy, Gandhi, Martin Luther King and others.

16 Compare, for instance, the Thoreau quote in this paragraph with King's: 'Noncooperation with evil is as much a moral obligation as is cooperation with good'; Martin Luther King (2000) [1958] 'Stride Toward Freedom', anthologised in Lauter, *Walden and Civil Disobedience*, p. 424.

17 Mohandas K. Gandhi (2000) [1920] 'Satyagraha', anthologised in Lauter, *Walden and Civil Disobedience*, p. 410.

18 Critical Art Ensemble, *Digital Resistance*, p. 14.

19 For CAE's response to FloodNet, see their *Digital Resistance*, chapter 1. There is also friction between CAE and their former member Ricardo Dominguez which seems to centre around ownership of the concept of electronic civil disobedience and its history. See, for example, CAE's open letter to Dominguez of 28 September 1998, archived at http://www.nettime.org.

20 William Gibson (1984) *Neuromancer*, HarperCollins, London, p. 80. It's interesting to compare the fictional Panther Moderns with the novelist William Burroughs's *non*-fiction theory of electronic revolution, through which a few activists with tape recorders could create chaos by playing back old recordings in new situations. See William Burroughs (1984) *The Job*, John Calder, London.

21 Dorothy Denning (2001) 'Cyberwarriors: Activists and Terrorists Turn to Cyberspace', *Harvard International Review*, vol. 23, no. 2, p. 72.

22 Niall McKay (1998) 'Pentagon Deflects Web Assault', *Wired News*, 10 September, http://www.wired.com/news/politics/0,1283,14931,00.html.

23 Winn Schwartau (2000) 'Cyber-Civil Disobedience: Inside the Electronic Disturbance Theater's Battle With the Pentagon', 23 February, http://www.infowar.com/chezwinn/articles022300/VigilSidebar2v3-pentagon1FinalDesmnd.shtml. Schwartau, we should note here, is not a fan of virtual sit-ins; he has been quoted as saying of participants: 'I think they're cowards. We never hid behind this veil of anonymity when we protested against Vietnam.' Noah Schachtman (2002) 'Rail Against Econ Forum, Dot-Org', (sic) *Wired News*, 30 January, http://www.wired.com/news/print/0,1294,50105,00.html.

24 Much of this coverage is archived in EDT member Carmin Karasic's online scrapbook: http://www.xensei.com/users/carmin/scrapbook/scrapbook.htm.

25 The fact that Mexico is marginalised within the US media also points to one response to a common objection to FloodNet-style denial-of-service attacks. Critics frequently call such actions an attack on free speech – if, say, the Pentagon website is being flooded in this way, then their right to distribute their own information is being denied. But what this argument overlooks is the question of who usually gets to speak. And who usually has their position framed for them by others?

26 Bertrand Russell faced the same question in relation to his involvement with civil disobedience in the Campaign for Nuclear Disarmament: 'By means of civil disobedience', wrote Russell, 'a certain kind of publicity becomes possible. What we do is reported, though as far as possible our reasons for what we do are not mentioned. The policy of suppressing our reasons, however, has only very partial success. Many people are roused to inquire into questions which they had been willing to ignore'; Bertrand Russell (1969) [1963] 'Civil Disobedience and the Threat of Nuclear Warfare', in Hugo Adam Bedau (ed.) *Civil Disobedience: Theory and Practice*, Pegasus, New York, pp. 153–59, quote from p. 157.

27 Quoted in 'Cyber Town Hopes for Web Success', *The Age*, 22 January 2000, p. 23. The town's own version of events is online at http://www.pinetel.com/~half/change.htm.

28 Amy Harmon (1998) '"Hacktivists" of All Persuasions Take Their Struggle to the Web', *New York Times*, 31 October, archived at http://www.xensei.com/users/carmin/scrapbook/scrapbook.htm.
29 John Arquilla and David Ronfeldt (1993) 'Cyberwar is Coming!', *Comparative Strategy*, vol. 12, no. 2, pp. 141–65, available online at http://www.rand.org/publications/MR/MR880.
30 Arquilla and Ronfeldt, *The Zapatista Social Netwar in Mexico*, p. xi.
31 Arquilla and Ronfeldt, 'Cyberwar is Coming!', p. 40. Arquilla has also contributed a short story (based on his research) to *Wired* magazine: John Arquilla (1998) 'The Great Cyberwar of 2002', *Wired* 6.02, February, http://www.wired.com/wired/archive/6.02/cyberwar.html. See also John Arquilla and David Ronfeldt (2001) 'Fighting the Network War', *Wired* 9.12, December, http://www.wired.com/wired/archive/9.12/netwar.html.
32 Critical Art Ensemble, *Electronic Civil Disobedience*, p. 18.
33 Philip Schlesinger, Graham Murdock and Philip Elliott (1983) *Televising Terrorism: Political Violence in Popular Culture*, Comedia, London, p. 1.
34 Schlesinger et al., *Televising Terrorism*, p. 2. On the high news value of official pronouncements, see also the 'propaganda model' of Edward S. Herman and Noam Chomsky (1988) *Manufacturing Consent: the Political Economy of the Mass Media*, Vintage, London.
35 The NSA text is at http://www.thing.net/~rdom/ecd/nsa_show1.html.
36 The line 'information wants to be free' is generally attributed to Stewart Brand. On its centrality to the hacker ethic see Levy *Hackers: Heroes of the Computer Revolution*, especially chapter 2. Perhaps an even purer expression of this now legendary anthropomorphism is John Perry Barlow's assertion that 'Information is a life form'. See Barlow (1994) 'The Economy of Ideas', *Wired* 2.03, March, http://www.wired.com/wired/ archive/2.03/economy.ideas.html.
37 Peacefire, http://www.peacefire.org. In classic hacker fashion, Peacefire's founder was just 17 when he started the site.
38 For a preliminary taxonomy of hacktivist tactics, see Dorothy Denning (1999) 'Activism, Hacktivism and Cyberterrorism: the Internet as a Tool for Influencing Foreign Policy', http://www.nautilus.org/info-policy/workshop/papers/denning.html, and Denning, 'Cyberwarriors'. See also the online zine *The Hacktivist*, http://thehacktivist.com.
39 Denning, 'Cyberwarriors', p. 74.
40 Mark McGrath (2001) 'The Crucifixion and Resurrection of Virtual Democracy', *Workers Online*, no. 100, 29 June, http://workers.labor.net.au/100/news22_virtual.html. See also Michael Bachelard (2001) 'Workers Party Online', *The Australian* Media supplement, June 28, p. 3.
41 Denning, 'Activism, Hacktivism and Cyberterrorism'.
42 Denning, 'Cyberwarriors', p. 70.
43 Kevin Anderson (2001) 'Hacktivists Take Sides in War', *BBC News*, 23 October, http://news.bbc.co.uk/hi/english/world/americas/newsid_1614000/1614927.stm.
44 Denning, 'Cyberwarriors', p. 71. Attrition is at http://www.attrition.org. *2600: the Hacker Quarterly* maintains an archive of before-and-after screen shots of hacked sites at http://www.2600.org/hacked_pages.

45 See http://www.s11.org and http://www.melbourne.indymedia.org for archived coverage.

46 Chris Alden (2001) 'Hackers Redirect Hamas Site to Porn', *The Guardian*, 7 March.

47 Stephanie Peatling (1999) 'Police on Track of the "Mad Hacker",' *Sydney Morning Herald*, 22 October, p. 6.

48 Deborah Radcliff (2000) 'Vigilante Group Targets Child Porn Sites', CNN.com, http://www1.cnn.com/2000/TECH/computing/01/11/condemned.org.idg, 11 January.

49 Denning, 'Activism, Hacktivism and Cyberterrorism'.

50 Karasic at http://www.xensei.com/users/carmin/scrapbook.

51 Levy, *Hackers: Heroes of the Computer Revolution*, p. ix.

52 Ibid., p. 10.

53 Sterling, *The Hacker Crackdown*, p. 274.

54 Douglas Thomas (2000) 'New Ways to Break the Law: Cybercrime and the Politics of Hacking', in David Gauntlett (ed.) *Web.Studies: Rewiring Media Studies for the Digital Age*, Arnold, London, pp. 202–11.

55 On the criminalisation of hacking, see Douglas Thomas (2000) 'Criminality on the Electronic Frontier', in Brian D. Loader and Douglas Thomas (eds) *Cybercrime*, Routledge, London, pp. 17–35. Thomas notes that in hacker trials, the charges often come down to *possession of computer equipment*, due to the difficulties of conceptualising non-physical 'trespass' or the 'theft' of information which was duplicated but not removed.

56 Philip Elmer-DeWitt (1988) 'Invasion of the Data Snatchers!' *Time*, 26 September, vol. 132, no. 13, pp. 62–67.

57 An interesting analysis of the Morris case is found in Andrew Ross (1991) 'Hacking Away at the Counterculture', in Andrew Ross and Constance Penley (eds) *Technoculture*, University of Minnesota Press, Minneapolis, pp. 107–34. See also Katie Hafner (1990) 'Morris Code: Computer Crimes and Misdemeanors', *The New Republic*, vol. 202, no. 8, 19 February, pp. 15–16.

58 Kevin Poulsen (1999) 'Exile.com', *Wired* 7.01, January, http://www.wired.com/wired/archive/7.01/poulsen.html.

59 Cited in Paul A. Taylor (1999) *Hackers: Crime in the Digital Sublime*, Routledge, London, p. 178; Taylor is quoting an unnamed participant in an episode of the *Geraldo* program. On the Mitnick case see the account by his eventual captor. Tsutomu Shimomura (1996) 'Catching Kevin', *Wired* 4.02, February, http://www.wired.com/wired/archive/4.02/catching.html. On the demonisation of hacking, see also Andrew Ross 'Hacking Away at the Counterculture', and Sterling, *The Hacker Crackdown*. Also of interest is Katie Hafner and John Markoff (1991) *Cyberpunk: Outlaws and Hackers on the Computer Frontier*, Corgi, London.

60 The demonisation of hackers can be usefully compared with that of the Luddites. The specific and considered responses of the Luddites in 1811 were counter to the emerging conditions of the industrial revolution – the triumph of the technocratic world view can be seen in the way the term 'Luddite' has been debased to connote an irrational opposition to the new. There was nothing

irrational about the Luddites. But they were out-rationalised. See E.P. Thompson (1980) *The Making of the English Working Class*, Victor Gollancz Ltd, London.

A different line of argument for future research might be to pursue the suggestion that social unease over hackers reflects a wider range of anxieties about the tensions between mastery of technology and/or its mastery over us. The significance of hackers here is both their impressive technical ingenuity and their courting of the consequences of challenging corporate control of technology. See Paul A. Taylor (2000) 'Hackers – Cyberpunks or Microserfs?', in Loader and Thomas (eds) *Cybercrime*, pp. 36–55.

61 The classic text on this topic is Stanley Cohen (1972) *Folk Devils and Moral Panics*, MacGibbon & Kee, London.

62 John Hartley (1992) *The Politics of Pictures*, Routledge, London, p. 207.

63 Sociologists offer competing theories as to what causes moral panics. Goode and Ben-Yahuda, for instance, in their survey of the field, identify distinctions between those panics that come from the top down (from political elites), those that come from the bottom up (from the grassroots general public), and those that come from special interest groups somewhere in the middle, including the media. One way into exploring a specific panic is to ask the question 'who benefits?' In the case of the hacker moral panic, we're spoiled for choice with answers. See Erich Goode and Nachman Ben-Yahuda (1994) *Moral Panics: the Social Construction of Deviance*, Blackwell, Oxford.

64 This explanation would be another instance of backing into the future – is computer intrusion, for example, really best described as trespass? Or as something completely new?

65 Thomas, 'New Ways to Break the Law', quote from p. 206.

66 Ross, 'Hacking Away at the Counterculture', quote from p. 111. Bruce Sterling points out that in the 1990 crackdown, the US Secret Service were largely ignorant of UNIX systems and were dependent for their understanding on the viewpoints of the very commercial operations who were crying foul. Operation Sundevil, the key event of the hacker crackdown, was, writes Sterling, 'greeted with joy by the security officers of the electronic business community.' See Sterling, *The Hacker Crackdown*, p. 163.

67 Sterling, *The Hacker Crackdown*, p. 57.

68 For background on the eToys/etoy case, see the following sites: etoy, http://www.etoy.com; Toywar, http://www.toywar.com; ®™ark, http://rtmark.com. A comprehensive archive of press coverage of the events is maintained at the ®™ark site. See also agent.NASDAQ (aka Reinhold Grether) (2001) 'How The Etoy Campaign Was Won: an Agent's Report', in Peter Weibel and Timothy Druckrey (eds) *net_condition: art and global media*, MIT Press, Cambridge, MA, pp. 280–85 (an earlier version of this text, 25 February 2000, is archived at http://www.nettime.org).

69 etoys closed its digital doors on 8 March 2001, the day after it filed for bankruptcy. Its name and website address were later purchased by another toy company.

70 Ellen Messmer (1999) 'etoys Attacks Show Need for Strong Web Defenses', CNN.com, 21 December, http://www.cnn.com/1999/TECH/computing/12/21/etoys.attack.idg/index.html.
71 Critical Art Ensemble, *Digital Resistance*, p. 14.
72 Ibid., p. 27.

Epilogue

1 Eszter Hargittai (2001) 'Radio's Lessons for the Internet', *Communications of the ACM*, vol. 43, no. 1, January, pp. 51–57.
2 One survey of 1500 US adults, for instance, conducted in late November 2001, found that 80% rated government and military censorship of news from the war in Afghanistan as a 'good idea'. Pew Research Centre for the People and the Press report online at http://www.people-press.org/112801rpt.htm.
3 Lawrence Lessig (1999) *Code: and Other Laws of Cyberspace*, Basic Books, New York, p. x.
4 Which is not to say that such activism doesn't already exist – it does. Some of the many sites of interest include: Electronic Frontier Foundation, http://www.eff.org; Index on Censorship http://www.indexonline.org; We Want Bandwidth!, http://www.waag.org/bandwidth; Fibreculture, http://www.fibreculture.org.

Index

For Freda and Ray